Guernsey Place Names

Guernsey Place Names

Guernsey Place Names

Hugh Lenfestey

RoperPenberthy Publishing

Map by courtesy of Digimap Ltd, Old Tobacco Factory, La Ramee,
St Peter Port, Guernsey GY1 2ET

RoperPenberthy Publishing Limited,
23 Oatlands Drive,
Weybridge KT13 9LZ,
United Kingdom

ISBN 978-1-903905-79-1

Jacket Design by Paul Stanier

Typeset by Avocet Typeset, Somerton, Somerset

Printed in the United Kingdom

Contents

John Hugh Lenfestey
1934–2012

John Hugh Lenfestey was known throughout his active life as Hugh to distinguish him from his father, John, and his grandfather, Jean. Born in Guernsey, Hugh was brought up at La Houguette, in the parish of St. Pierre du Bois, where the family have lived for over 250 years.

At the age of six he evacuated with his mother and siblings to England, ahead of the German Occupation of the Channel Islands in 1940, whilst his father remained, looking after the family properties. Following the Liberation in 1945, Hugh continued his education, initially at Elizabeth College in St. Peter Port and then at Southampton University where he read economics. After graduating, Hugh worked at the Greffe (Guernsey's Record Office) with Dr Conway Davies of the then English Public Records Office (now the National Archives) cataloguing the Island's collection of historical documents. Many of these irreplaceable sources dated from the Middle Ages and were written in contemporary and ancient legal French.

Hugh worked on this project for three years until the States of Guernsey (the Island government) discontinued its funding. He then obtained a Certificate in Education from Bristol University and after teaching in England for six years returned to Guernsey. He taught in primary education for the next twenty-five years.

In 1985 the States of Guernsey decided to create an Island Archives Department to manage and conserve the Island's governmental papers and its unique collection of medieval documents. Hugh was appointed Island Archivist and also *Archiviste de la Cour Royale* (Archivist to the Royal Court of Guernsey), holding these important posts until his retirement in 1997.

Hugh had developed a deep and abiding interest in local history at a young age. Given a bicycle at the age of 16 to celebrate passing his School Certificate, he began to explore the Island outside his home parish. Engaging the older generations in conversation he became aware of the feudal system which was still in active use in the late

1940s. He learnt the names of local areas and of specific fields, and also why they were so named, and what the names meant.

Although still relatively young, Hugh had the rare foresight to gather knowledge and empirical evidence about a rapidly passing historical age. Whilst Guernsey had endured the German Occupation, it now faced an increasing British "occupation" from settlers across the other side of the English Channel. In the following decades this second "occupation" tended to dilute the culture and identity of the Guernsey people. Without Hugh's robust efforts, most of Guernsey's unique heritage would have been irretrievably lost forever. What started originally as a school boy hobby became in time a body of systematic world class research.

Hugh's grandfather Jean Lenfestey, the Seigneur of Fief des Philippes, encouraged Hugh in this interest and taught him how to read and translate the *livres de perchage* (books of measurement) which list the properties within each feudal holding, and which were usually produced once in each generation.

Hugh began to collect these *livres de perchage* and over his lifetime acquired a unique collection of over one thousand books, both original and copies, comprising all the extant *livres de perchage.* In addition he also put together an extensive collection of historical documents and conveyances, some dating from the early 1500s, and it is these collections that have formed the basis for his list of place names. After a lifetime of research, Hugh finally completed the manuscript for this book shortly before his untimely death.

This book is particularly important now when there is an increasing interest in dialect languages and Hugh's work on derivations and identifications will help to correct misunderstandings about the origins of local names.

The names in this index are those that can be proved to apply to the area described. Many of the names can be translated into English, but where the derivation is in doubt they have been left in the original French or Guernsey-French. There are topographical lists of the most commonly used words and their meanings and an introduction by Hugh explaining what the book is about and the way in which he wrote it.

Regarding his own surname, Hugh was able to confirm that the name l'Enfete was one of the original names given to Guernseymen in the early 12th century. It was so written, as *L'enfete*, or *L'enfeste*, until the beginning of the 19th century. Only the increasing numbers of English people in the Island have caused the eventual Anglicisation of the name to Lenfestey.

Hugh maintained a unique historical perspective of the culture and place of Guernsey, sitting as it does between France and Britain.

Once, on being told that "things on the Mainland were not going too well economically", Hugh characteristically replied, "What, France?".

Although Hugh pursued a deeply academic life and was a man of immense erudition, he wore his learning lightly. In true Lenfestey fashion, Hugh turned his hand to anything practical and enjoyed driving large agricultural implements and helping fellow farmers on the Island.

His family are delighted to see his work published and are grateful to all those who have made it possible.

Gillian Lenfestey
Les Adams de Haut
St. Pierre du Bois
Guernsey

Introduction

Most of the place-names in Guernsey are very old and their roots often go back to the late medieval period. Many of them are topographical and often have either a forename or a surname of an individual attached to them.

Place-names that are derived solely from family names proliferated considerably from the 13th century onwards, following the general development of hereditary surnames in the later 12th century. There are some exceptions which are derived from past associations with legal titles, arising from rentes or 'saisis' (judicial seizures) of property. Nevertheless there are a few which are not explicable or lack conclusive proof of their origins, either because their form has changed out of recognition or the historical roots can no longer be traced.

Surviving feudal records and conveyancing documents of the 15th and 16th centuries show that each district of houses and land in the island had not only a distinctive name, but often included detailed comprehensive descriptions for the smallest areas of land and their boundaries, whether enclosed or unenclosed

There was clearly a need for such exactitude, which arose from the considerable burden of perpetual mortgages (rentes) and feudal dues (chefrentes and champarts) owed on all real property in the Island up to recent times.

This is why the detailed extent of the properties on each fief was renewed and updated once in every generation, the books being called 'livres de perchage'.

It is worth noting that houses traditionally carried the name of the immediate locality, rather than having individual names. Only roads with specific functions or particular features were named, with a few exceptions. For example we have le chemin le Roi (king's highway), la Grande rue (main road), la rue Poudreuse (dusty road) and rue au Ferré (metalled road).

Some main roads such as the Route de la Ville à l'Erée, and the Route de la Ville à Vazon were named in the 19th century for their entire length from St Peter Port, but by the 20th century each section of these roads had its local district place-name. More further naming of roads was carried out in the 1970s and 1980s to provide localities

Camps de terre strip field

with more adequate addresses, but unfortunately that process has been inconsistently applied and has only led to a further confusion of place names.

Half of these roads carry the name of the district, i.e. rue du Préel and route des Prevosts, and can produce conundrums such as "Les Martins Villa, rue des Martins, Les Martins, St Martin's". The other half has the name of the district to which the road is going, i.e. route de Jerbourg and route de Côbo, which is often wrongly interpreted and that place-name is then 'stretched' a considerable distance before it reaches its actual location.

The majority of the place-names given in this Index arose from a highly localised subsistence agriculture practised everywhere in the island parishes outside the town of St Peter Port. Parallel with that there is also a body of place-names of considerable antiquity; trépied, pouquelaie, castel, déhus, déhuzet, pierre, rocque, which arose from an ignorance of archaeology and historical method in the middle ages, and instead gave rise to legends and folk tales.

The written language in island records changed from Latin to French early in the 14th century. There are, therefore, very few recorded instances of Latin place names in the island, and these have been gallicized. The written French corresponds to that used in similar legal and quasi-legal documents in Normandy for the late medieval to early modern historical periods. This has led to the survival of topographical descriptions and in many areas names of expressions that have become either obsolete or lost in the modern French usage of the 19th and 20th centuries.

In addition, various contracted spellings and the occasional misspelling from a particular pronunciation have arisen from the method of dictation used over the past 300 years to 'speed write', in order to more quickly duplicate most documents and records. This has happened with all classes of records, from those of the Royal Court, feudal and parish records to those of family inventories of personal possessions.

Other factors have included the development of plural forms to cover districts where there was originally only one building, but over time there is now more than one house or field of a particular name, whether a family surname or a topographical feature. Differences in the local pronunciation in Guernsey French has also led to variations in spelling, but the lack of an agreed orthography has made it difficult to successfully reproduce place names phonetically to safeguard that pronunciation for posterity.

No attempt is made in this Index to consider the etymology of the various place names unless, as in a very few cases, that has a bearing on the use of a place name for a particular locality. It would appear nonetheless that most Guernsey place-names have Norman roots, presumably partly derived from the language of the early medieval Norman settlers in Normandy.

During the 20th century Guernsey became increasingly suburbanised in all but a few areas, and the need for precise place-names even within the more rural localities has disappeared. Even the farming community has lost interest, and by losing so many once common field names most farmers now have difficulty in identifying their fields by any name. Some have resorted to numbering them, or even using their cadastre* numbers. This rapid disappearance of place-names from everyday use has been very considerably influenced by the universal introduction of the English language since the Occupation.

There is now an 'English' outlook to Guernsey place-names, with the increasing introduction of an English, rather than even a Guernsey English, pronunciation, for example Port**in**fer for Portin**fer** and Pal**lot**erie for Pallot**erie**. There is an increasingly 'English' emphasis on the first or middle syllable of a place-name, rather than giving a strong inflexion to the ending. However it must be said that there has been a long Guernsey tradition of 'clipping' a name whenever possible, and this has also contributed to the present situation.

A further development has been the creation of 'village' names in place of well known district names. Hence 'St Martin's Village' for La

---

* land survey.

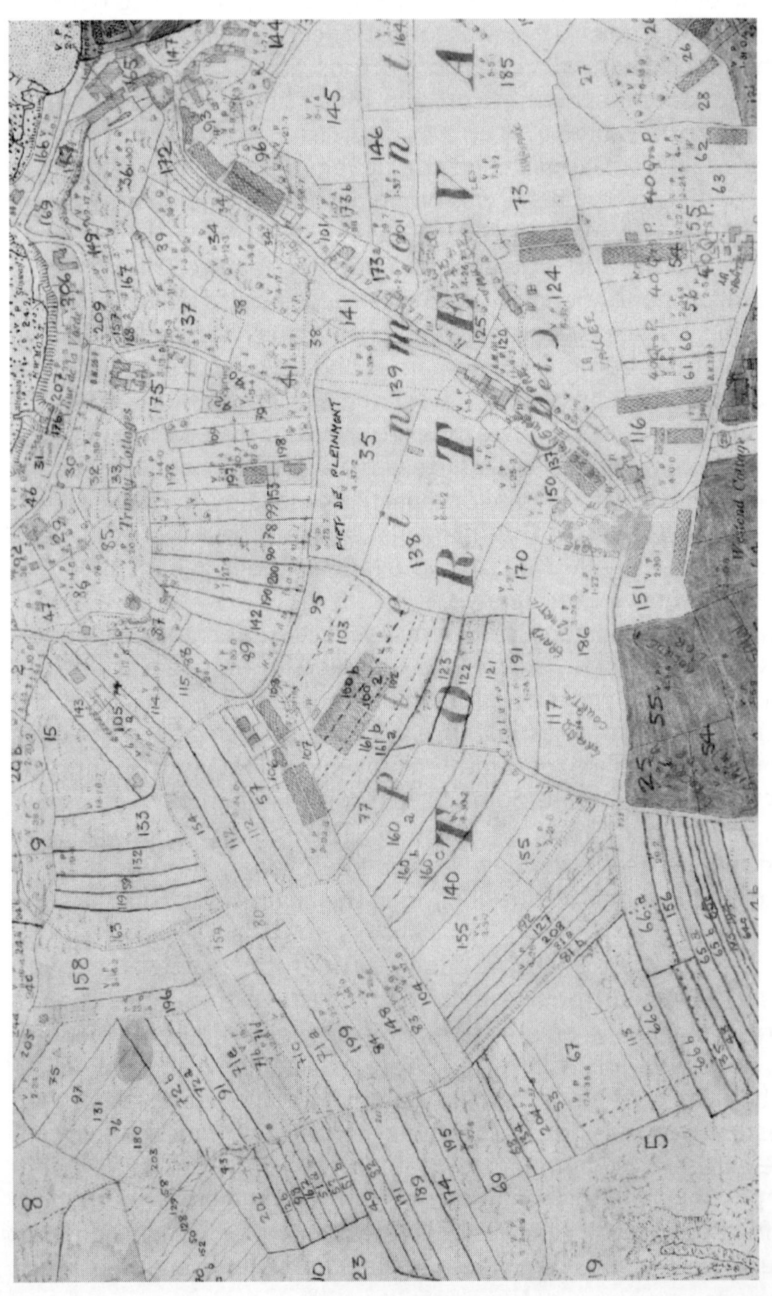

Pleinmont map

Bellieuse and 'St Peter's Village' for Les Brehauts and Les Buttes. In the latter case, Les Buttes have even become the 'village green'. No attempt has been made to include these "bastard" names and the more recent names for houses and estates, as these tend to be either historically distorted or totally whimsical.

The situation has now been reached in many parishes where one place-name will 'do' for a district, covering an area within a half-mile or so of its original location. A result of this is that most of the place-names in this Index have become 'forgotten' especially during the last fifty years, and those in use have become 'labels' for parts of the Island.

Hence, for example, L'Islet which was formerly an islet (now called Sandy Hook) on the edge of the former Braye du Valle has been stretched to cover an area a square mile westwards. It now encompasses Les Petites Mielles (in itself a large area), la Maladerie, Rungefer, Route Carré, les Romains, les Trachieries, les Frocs, Picquerel, les Marettes, le Clos, les Salines, Rue Colin, les Martins and la Grande Maison.

Other examples of 'stretched' place-names include Pleinmont, Torteval; l'Erée and le Coudré in St Pierre du Bois and Talbot Valley in the Castel and St Andrew.

Use of family surnames

The place names derived from family names, many of which have been long since extinct, have developed distinctive forms of their own. These fall into eight groups, of which the following are examples.

1. Use of the original form
Sohier, Lorier, Roberge from the surnames Sohier, Lorier, Roberge.

2. Use of the original in a 'clipped' form
Retôt, Becrel, Rignet from the surnames Restault, Becquerel, Riguenet.

3. Use of a simple plural form
les Eturs, les Fauconnaires, les Frances from the surnames Etur, Fauconnaire, De France.

4. Use of the plural form with various suffixes
les Duvaux, les Grippios, les Hubits, les Raies, les Emrais, from the

surnames Du Val, Gripel, Le Huby, Le Ray, Emerey.

5. Use of an accentuated form
les Cambrées, les Grantez, les Rouvets, les Villets, from the surnames Cambre, Grente, Le Rouf, De Ville.

6. Addition of the suffix 'erie'
la Bertozerie, la Cailleterie, l'Etonnellerie, la Pelleyrie from the surnames Bertault, Caillet, Lestournel, Le Pelley.

7. Addition of the suffix 'ière'
la Besnardière, la Heronnière, la Plichonnière from the surnames Bernard, Le Heron, Pelichon.

8. A significant change in form
le Gelé, les Brayes, les Houards, la Pomare, les Queritez, les Vicheries from the surnames Le Gelley, Le Bret, Ouar, D'Appemare, Caritey, Le Vecheulx.

Glossary of Topographical Expressions

A Glossary of Topographical Expressions will be found after this Introduction. It is intended to be a simple guide to the considerable number of these expressions. Most of them have fallen out of use and are not easy to source.

Topographical and other place-names

Topographical names are generally fairly easy to recognise, and within this group are also included the names given to the various types of megaliths and other man made structures. Examples of these include la Pouquelaye, la Longue Rocque and le Câtioroc (Câtel au Roc). There are also surnames derived from the topography itself, i.e. De Lande, De Fosse, Du Douit, Du Val, Du Mont, De La Mare, Du Maresq, Du Pré and Des Mares.

Topographical names are also combined with family surnames, hence La Houguette au Tillier, le Valangot (Val Angot), les Vauxbelets (Vaux Belic). Agricultural use has also provided examples, le Camposo (Camp Oso), les Champhuets (Camp de M. Hue).

A final category of place-names contains the miscellanea of names associated with religious bodies receiving rentes, possible nicknames and the use of forenames. Firstly there are Senicolas for the confraternity of St Nicolas, and Nôtre Dame, for a rente to 'Our Lady'. Secondly, les Plats Pieds may have been an 18th century nickname, while two houses at les Rouvets, Vingtaine de l'Epine (listed in the Index) had informal nicknames during the 19th century, one being 'Moutonnerie' and the other 'Côte ès Ouets'. Thirdly, a forename was very occasionally used, thus les Laurens from Lorens, Perotin (from Pierre), Lorenche (from Laurence), Colin and Colliche (from Nicolas), with Benoit being used as both a forename and a surname.

Abbreviations

Parish names, with the two divisions of the Vale:

SPP	St Pierre Port	St Peter Port
SSP	St Samson	St Sampson
CduV	Clos du Valle	Vale (N. of le Braye du Valle)
VdelE	Vingtaine de l'Epine	Vale (on N. of Castel)
C	Castel	Castel
SSV	St Sauveur	St Saviour
SPB	St Pierre du Bois	St Peter in the Wood
T	Torteval	Torteval
F	Forêt	Forest
SM	St Martin	St Martin
SA	St André	St Andrew

Expressions: f = feminine. m = masculine

Notes regarding the Index

Place-names are listed under the more generally used spelling, and cross references are made to alternatives and noteworthy historical citations. It should be noted that in order to retain their historical context many of the references used relate to 'fields' which in several parishes no longer exist and instead are now part of a larger suburban landscape.

A capital 'D' and 'L' has been used for all surnames beginning with 'de', 'du', 'des' and 'le' to show that these are an integral part of that name and are not grammatical expressions relating to their origins.

Locations are given for the parish or parishes in which the various place-names occur with the district, as a single entry, for each property within that parish entry.

All the feudal holdings except those on the Clos du Valle have been mapped and digitised, and are in the process of being labelled, which will give a more precise location for the individual areas.

Group lists are given for bordages, camps, chapelles, clos, courtils, croix, croûtes, surnames with De, douits, écluses, fiefs, fontaines, forts, fosses, fossés, names with grand(e), hèches, hougues, houguettes, hures, surnames with Le, landes, manoirs, mares, marais, ménages, monts, rues, ruettes, val and vaux, vallées, vallettes and villes. These are intended to be as complete as possible, allowing for the historical sources available. The individual place-names under each heading are also to be found in the main text. In addition archaeological references to 'megaliths' are noted under castel, chaise, creux, déhus, déhuzet, perron, pouquelaie, pierre, rocque and trépied.

Measurement of land

Both Guernsey and Jersey have land measurement systems which are now unique to them, which are of Norman extraction. They were vital for use in exploiting both large and small areas of land. Now that local agriculture in both islands is very much a minority occupation employing a small number of people the need for such a detailed measurement system has nearly disappeared. English and French measures are now used more frequently, which helps to show how cosmopolitan the Islands have become during the last 50 years.

The largest unit of superficial land measure, that is square measurement, was the bouvée and although long since obsolete and out of use it figures in many place names as a district name, often in combination with a now extinct family surname. The bouvée contained 20 vergées, and the vergée still remains our principal measure of land. The acre of four vergées has long been out of use except as a place-name, and the English acre is increasingly being used in place of the vergée. The vergée divides further into perches, 40 perches grande mesure (big measure) which is the usual division, or 36 perches petite mesure (small measure) which became obsolete in medieval times.

The exchange rate between Guernsey and English superficial measure is recognised as being 2½ vergées to an English acre. There is in fact a discrepancy of some 3 square inches one way or the other, which can be ignored for ease of comparison.

The local linear measurement of land has died out. It was the verge of 7 yards, which happened to be the length of one side of the perch noted above. It has long been replaced by the ordinary English yard and since the 1980s by the metre.

Sources of information

This work only includes place-names that have at least one reasonably authentic reference as a place-name. This is why of the holdings of the various feudal offices which had perquisites only a very limited number have been included, and these mainly relate to bordages held by some bordiers and prévôtés by a few prévôts.

Derivations of place-names are given where they can be worked out satisfactorily, otherwise the comment is made suggesting that a probable or possible meaning can be deduced. Failing that, the origin is said to be unknown. None of this is helped by spellings arising from dictated documents, and the local habit where possible of contracting spellings and pronunciations into two syllables from an original three or even four-syllabic word.

Principal references

Greffe Catalogue, Volumes 1, 2 and 3.
Document collections, Island Archives Service, Guernsey.
Livres de perchage of island Fiefs, which cover most of the island from the 15th century to the mid-20th century.
Cartulaire des Iles Normandes. pub. La Société Jersiaise. 1919–1924.
The Extentes of Guernsey, 1248 and 1331. Sir H de Sausmarez, editor. La Société Guernesiaise,1934.
Extente des Iles de Jersey, Guernesey, Aurigny et Serk, 1274. pub. by La Société Jersiaise, 1877.
Dictiounnaire Angllais-Guernesiaise. Marie de Garis. 2nd Edition 1982.
Dictionnaire Jersiaise-Français. Frank Le Maistre. 1966.
Glossary of Guernsey Place names. Marie de Garis. 1976.
Guernsey Street and Road Names. Compiled by C J Howlett. 1983.
Jersey Place names. 2 vols. C. Stevens, J. Arthur, J. Stevens. Revised and supplemented, 1986.
Alderney Place Names. Royston Raymond. 1999.

Thanks

My sincere thanks go to my wife for typing out and proofreading the manuscript, and to Dr D M Ogier and Mr W T Gallienne, who have painstakingly read the unedited text and have greatly improved it. Nevertheless errors of fact and understanding are mine only.

Hugh Lenfestey
January 2012.

Topographical Descriptions

French to English

Abreuveur (m), abreuvoir (m)	trough, watering place
Allée (f)	private impermanent crossing on a slope, capable of being grazed
Assiette (f)	site (of building)
Banc (m), banque (f)	natural coastal bank
Barrat (m)	ditch
Béguin (m)	beacon on high ground
Belle (m)	paved yard, adjacent to buildings
Bin (m)	sluice for mill pond
Borde (f)	unenclosed plot bordering road
Braye (m)	tidal straits between islands
Brayes (m. pl.)	name given to shoal of rocks in maintidal stream
Butte (f)	butt (for crossbows)
Cache (f)	private cart track, with banks on both sides (ownership separate from adjoining land)
Camp de terre (m)	strip of land
Carrefour (m)	crossroads
Carrière (f)	1. alt. spelling 'charriere' (f) q.v. 2. quarry, only in 19th /20th centuries
Chapelle (f)	chapel
Charrière (f) [c.f. querrière (f)]	private permanent stony track for carts, without banks
Château (m)	castle
Chemin (m), route (f), rue (f), ruette (f) voie (f)	highway: major road: road : lane: way
Cimetière (m)	cemetery
Clos (m), Clôture (f)	enclosure
Courtil (m)	field
Courtillet (m)	small field
Croix (f)	(wayside) cross
Cure (f)	living, of ecclesiastical parish

Dégrais (m)	steps, (public thoroughfares)
Douit (m)	stream
Écluse (f)	mill pond
École (f)	school
Église (f)	church
Entrée (f)	entry into property
Fallaize (f)	cliff
Fallaisette (f)	low cliff, steep slope
Feugré (m), feugrel (m), feuguerel (m)	bracken brake
Fontaine (f)	spring
Fosse (f)	pit, ditch
Fossé (m)	artificial field bank
Fouillage (m)	'fuel' brake (bracken/gorse)
Fourrière (f)	kiln
Frie (m)	unenclosed land
Friquet (m)	diminutive of above
Frichet (m)	alternative of friquet
Garenne (f)	rabbit warren
Gensage (m)	lay-by
Grange (f)	barn
Hannière (f)	area of 'han', (sedge)
Hautgard (m)	rickyard
Hâvre (m), port (m)	harbour, port
Heaume (m)	islet
Hommel, hommet (m),	diminutive of heaume, small islet
Hommetel (m)	islet, diminutive of above
Hougue (f)	hillock, on a spur
Houguette (f)	hillock, diminutive of above
Hure (f)	ridge
Hurel (m)	hollow, set in a ridge
Issue (f)	exit from property
Jaonnière (f)	furze or gorse brake
Jardin (m)	garden
Jardinet (m)	small garden, diminutive of above
Lande (f)	heath, wasteland
Largisse (f)	unenclosed strip bordering road
Maisière (f), mézière (f)	croft, small holding.
Maison (f)	house
Maisonnette (f)	cottage
Maison de garde (f)	watch house
Marais (m)	marsh
Maresq (m)	marsh, (older spelling)

Marché (m)	market
Mare (f)	natural pond
Marette (f)	pond, diminutive of above
Ménage (m)	paddock, used by the farm
Mont (m)	promontory
Monterel (m)	promontory, diminutive of above
Moulin à eau (m)	water mill
Moulin à vent (m)	wind mill
Mur (m)	rampart, wall
Muraille (f)	stone wall
Parc (m)	pig enclosure
Parquet (m)	diminutive of above
Paroisse (f)	parish
Pignon (m)	gable of building
Pointe (f)	unenclosed point of land
Porte (f)	principal doorway
Portière (f)	garden in front of the door
Pré (m)	meadow
Presbytère (m)	rectory
Prieurié (m)	priory
Puits (m)	well
Querrière (f), [c.f. charrière (f)]	permanent stony track for carts, without banks, across private property
Raulle	an animal pound
Rocque (f), pierre (m)	rock mass, standing stone
Rocquette (f)	diminutive of above
Saudrée (f)	willow patch
Séchage (m), sécheur (m)	varech (vraic) drying area
Tour (f)	tower
Terrain (m), terre (f)	terrain, land
Tertre (m)	hill
Travers, à (m)	archaic form of traversain
Traversain (m)	a 'crossing', i.e. a private, impermanent crossing over land, without banks, capable of being ploughed
Val (m)	valley
Valle (m)	vale
Vallée (f)	valley terrace
Vallette (f)	valley terrace, diminutive of above
Varde (f)	steep height
Varende (f)	rabbit warren

Vivier (m)	artificial pond
Voie (f)	way (misspelt in early 19th century as 'vue')

SURNAMES derived from the above:

Belle
Chemin (Quemin), Du
Du Douit
Fallaize
De Fosse
Heaume
Hommet (Houmet)
Des Mares
De Lande
Du Maresq
De La Mare
Du Mont
Moullin
Du Parcq
Du Pre
Du Val
Vivier

Topographical Descriptions

English to French

Bank, of field	fossé (m)
Bracken brake	feugré (m), feuguerel (f)
Castle	château (m)
Cemetery	cimètiere (m)
Chapel	chapelle (f)
Church	église (f)
Cliff	fallaize (f)
Croft, small holding	mézière (f)
Cross (wayside)	croix (f)
Cross roads	carrefour (m)
Ditch	fosse (f)
Enclosure	clos (m), clôture (f)
Field	courtil (m)
Furze brake	jaonnière (f)
Garden	jardin (m)
Harbour/port	havre (m), port (m)
Hill	hougue (f), tertre (m)
Height	varde (f)
House	maison (f)
Islet	hommel (m), hommet (m)
Living (of parish)	cure (f)
Market	marché (m)
Marsh	marais (m)
Meadow	pré (m)
Megaliths	chaise (f), perron (m), pierre (f), rocque (f), câtel (m), creux (m), déhus (m), déhuzel (m), pouquelaie (f), trappe (f), trépied (m)
Mill	moulin (m)
Mill pond	ecluse (f)
Paddock	ménage (m)
Parish	paroisse (f)
Pond (natural)	mare (f)

Pond (artificial)	vivier (m)
Pond (mill)	écluse (f)
Priory	prieurie (m)
Promontory	mont (m)
Rectory	presbytère (m)
Rickyard	hautgard (m)
Ridge	hure (f)
Road / lane / way	chemin (m), route (f), rue (f), ruette (f), voie (f)
Rock	rocque (f)
School	école (f)
Spring	fontaine (f)
Straits	braye (m)
Stream	douit (m)
Strip of land	camp de terre (m)
Tower	tour (f)
Track	cache (f), charrière (f), querrière (f), traversain (m)
Valley	val (m)
Valley terrace	vallée (f)
Wasteland	lande (f)
Warren	garenne (f)
Watch house	maison de garde (f)
Well	puits (m)
Yard	belle (m)

Places A–Z

A

Abbaye, l' — land: adjacent to priory church of St Michel-du-Valle, CduV. Late of abbey of Mont St-Michel, Normandy. Seized by the Crown by late 14th c. as land of an alien priory.

Abbesse de Caen, Fief de l' — feudal holding of land. Formerly belonged to a French religious house, now escheated to the Crown. SA.

Abraham Gallienne, hougue d' — land: St Briocq, SPB.

Abrégeas, les — fields: les Houmets, C; la Gallie, SPB. 1661. Also spelt **Abrégis**. Surname extinct 16th c. **Brégeart**.

Abreuveurs, les — houses, land: S of les Capelles. SSP. Property adjacent tol'abreuvoir (cattle watering place).

Abreuvoir, l' — cattle watering place, at a **douit** (stream) or **fontaine** (spring), all parishes. Also spelt **abreuveur**.

Acbuisson, l' — fields: S & W of le Clos Landais, SSV/SPB. Also spelt **Albuisson**, **Apbuisson**. Land formerly in 'white thorn'. Bissonerie, fosse q.v.

Accusson, l' — see **Cuchon**, le.

Acier, d' — field: le Moulin-de-haut, C. Origin unknown. Possible association with bee hives.

Acre (m) — Guernsey acre (4 vergées of land).

Acres, les **Quatre**	fields: E of la Cherverie, C.
Acuchon, l'	see **Cuchon**, le.
Adams, les	district between la Houguette and les Sablons, SPB. From surname extinct in Guernsey about 15th c.
Adoubez, croute des	land: la Ville Amphrey, SM.
Adoubez, les	land: les Camps-du-Moulin, SM. Presumably from extinct surname.
Agneaux, des	field: les Paysans, SPB; le Russel, F. Origin unknown, may relate to sheep-fold. Recorded: **Ancgneaulx** 1504, **Anniaulx** 1633, **Angnios** 1671.
Ahier, d'	field: les Poidevins, SA. Extinct surname, **Ahier**.
Aiguillon, l'	cliff-land: Jerbourg, SM. 'Pointed rock, needle shaped'.
Aitte, de l'	see **Laitte**, de.
Aix, ès	fields: le Frie, SPB; les Jehans, T; la Ville-Amphrey, SM. 'garlic'.
Albecq	field, district: (unloc) SPP; W of la Lande, C. Surname extinct 17th c. **Dallebec** (**Rauf Dalbec**, 1639).
Alebec, le **Maresc**	marsh: W of les Goddards, C. Land held half and half by abbey of Mont St-Michel and seigneur of Fief le Comte. Derivation as above. Fleurie Dalbec q.v. Also feudal holding of land, Fief **d'Allebec**, T.
Alexandre(s), les	fields: le Houmtel, CduV; N of le Lorier, SSV. Surname extinct 20th c. **Alexandre** (extant Jersey).
Alichette, d'	field: le Douit, SPB. Forename (f) **Alichette**.
Allaire, de **Jean**	field: le Neuf Clos, SSV. Surname extinct 20th c. **Allaire**.
Allaire, le **Pont**	former causeway: le Braye-du-Valle at la

	Ville-Baudu, CduV. Recorded 1591. Derivation as above. Pont St Michel q.v.
Allart, de	field: les Guelles, SPP. Origin unknown.
Allée (f)	way across a slope, capable of being grazed.
Allée, l'	land: slopes below St Saviour's church, SSV; garden, slope at le Manoir, F; slope W of la Marette, SM. Definition as above.
Allées, les	land: slopes off George road, SPP; slopes at les Choffins, rue Feveresse, and near les Bordages, SSV. Definition as above.
Alleris, les	field: le Gron, SSV. Origin unknown.
Alleu, l'	land: rue de l'Eglise, SSP; house & land, near Torteval church,T. 17th c. spelling la **Leur**. Also spelt **Alleur**. Freehold land, glebe belonging to the living (la **Cure**) of the parish.
Allez, de **Catherine**	field: les Caches, C. Extant surname, **Allez**.
Allez, de **Guillaume**	field: le Belle, SPB. Derivation as above.
Allez, la **Fosse Robert**	field: les Fosses, F. Derivation as above.
Allez, le **Bordage**	feudal holding of land, N of rue des Delisles, C. Derivation as above. Bordage q.v.
Allez, rue	1. early 19th c. street, Neuve Ville (Newtown), SPP. After the contractor, **Jean Allez**. 2. early 19th c. name for E end of Route du Braye, CduV. After the above **Jean Allez**, one of the four parties in its development. Braye-du-Valle q.v.
Allix de Lande, clos	land: unloc. CduV.
Allonné, l'	field: E of les Arguilliers, VdelE. Surname extinct 16th c. **De Laune**.
Almenac, le **Bordage**	land: N of les Grands Courtils, SA. Extinct surname **Amena** (Extente 1331).

Amballes, les	district: SPP. Surname extinct 19th c. **Lamballe (Francis Lamballe**, butcher).
Amelynez, ès	land: rue des Grons, SM. Surname extinct 16th c. **Ammellyne. Guillaume Ammelleyne**, 1488.
Amende, le **Val** d'	field: N of la Hure Godfrey, SPB. 'penalty' presumably once owed on a field. val q.v.
Amherst road	road: les Guelles, SPP. Early 19th c. road and barracks named after Earl Amherst, Governor of Guernsey 1770–1797.
Ammarreurs, les	'moorings': le Grand Havre, CduV.
Amont, la **Pointe** d'	'steep point of land'.
Amphrey, la **Ville**	see **Ville Amphrey**, la.
Ampière, d'	see **Dan Pierre**, de.
Ancresse, la **Baie** de l'	bay: CduV. 'anchorage'.
Ancresse, les **Communes** de l'	commons: l'Ancresse, CduV. communes q.v.
André, la **Fosse**	district: SPP. Surname extinct 14th c. **Landry**. fosse q.v.
Ane, la **Rocque** à l'	see **Rocque** à l'**Ane**.
Ane, le **Dos** d'	see **Dos** d'**Ane**, le.
Ange, la **Fosse** à l'	land: les Fontenelles, SSV. 'monkfish'.
Ange, la **Mare** à l'	land: les Mares, SPB. Derivation as above. fosse, mare q.v.
Ange, **Rue** à l'	cul de sac: off route du Friquet, C.
Angelots, les	land: E of ruette Forest (Forest Lane) SPP. From extinct surname **Angerlot**. Recorded **Angelotz**, 1574.
Angle, l'	fields: la Hougue Patris, CduV; le Friquet, C. Surname extinct 14th c. **Angle (Jordan Laungle**, 1309). rue q.v.
Angle, rue de l'	road: W of le Friquet, C.

Angot, la **Fontaine**	land: les Sauvagées, SSP. Surname extinct 14th c. **Angot**.
Angot, le **Val**	fields: S of les Portelettes,T; N of le Mont Herault, SPB. (**Guillaume Angot**, 1488). Derivation as above. fontaine, Valangot, q.v.
Anguillières, les	shingle bank: l'Erée, SPB. 18th c place name. 'place of eels'. Earlier name le **Gallet** de la **Rousse Mare**. gallet, mare, rousse q.v.
Anley, d'	field: Grange, SPP. Surname extinct end of 18th c. **Anley**. (Diary of Elisha Dobrée).
Ann's place	1. house: 'St Anns' Place', built 18th c., part of Old Government House Hotel. 2. road: early 19th c. road name, '**Ann's Place**', original extent from mid-Rue des Forges to Rue du College. Part of the former la Chasse Vassal q.v.
Anne, de **Ste**	houses, fields: King's Mills, C. Formerly part of **Manoir de Ste Anne** q.v.
Anne Jouanne, croute d'	land: unlocated, CduV.
Anne, le **Pont**	fields: les Landes, SM. Origin uncertain. Recorded, le **Pont Hané**, 1685.
Anne Nicolle, d'	see **Nicolle**.
Anne Nicolle, la **Croix** d'	see **Revel**, **Croix Dan** (Dominus) **Nicolas**. croix q.v.
Annevilles, les	district: SSP. Surname extinct 13th c. **D'Anneville**. Garenne D'Anneville q.v. Also feudal holding of land, Fief **D'Anneville**, SSP, from grant to **Sampson D'Anneville**, mid-late 12th c.
Annevilles, les	house: N of les Padins, SSV. 18th c. house name.
Annevilles, rue des	see Annevilles.
Anquetil, d'**Edouart**	field: le Neuf Chemin, SSV.

Anquetils, les — fields: W of la Hougue Renouf, C; la Rue Frairie, SA. Surname extinct 16th c. **Anquetil**. The current family is French.

Anthan, la **Hougue** — fields: S of Rue de Laitte, SPB. Surname extinct 15th c. Anthan, hougue q.v.

Antoine, d' — land: St Clair, SSP. Forename (m) **Antoine**.

Apbuisson, fosse d' — land: le Clos Landais, SSV. Also spelt Acbuisson.

Appemare, d' — see **Pomare**, la.

Appoline, **Ste**. — see **Ste Appoline**, **Chapelle**.

Arcade(s) —

1. les Arcades (States Arcades) part of mid-19th century market building. SPP
2. Commercial Arcade, early 19th century development by Le Bouteillier, quarrying part of le Mont Gibel q.v., between High Street (la Grande Rue) and the market (le Marché). SPP.

Archiers, les — house, fields: E of la Camp Tréhard, les Bémonts, SA. Surname extinct 15th c. **Larchier** (**Jean Larchière**). Arquet q.v.

Arculon, l' — district: le Bouet, SPP. house name: 17th c. probably from 'à reculons', inferring a 'back facing' or 'isolated' location.

Ardaines, les — fields: W of le Frie, SPB. Surname extinct 15th c. **Lardaine**.

Ardaines, rue des — see **Ardaines**.

Ardel, le **Mont** — field: les Osmonts, SSP. Extinct surname **Ardel**. mont q.v.

Arg(u) **ille**, la **Fosse** à l' — slopes: Havelet, SPP; le Moulin-de-haut, Woodlands, les Genâts, C. 'claypit'.

Arguilliers, les — district: N of le Douit, VdelE; rue du Val, SPB. Surname extinct 15th c. **Arguillier**. Also feudal holding of land, Bouvée des **Arguilliers**, VdelE.

Arguilliers, rue des — see **Arguilliers**.

Arquet, l' — fields: S of les Rouvets, SSV/SPB. Probably from above surname **Larchier**. Archiers q.v.

Arquet, rue de l' — road: late 16th c. 'Traversain' recognised as rue in 17th c. with hard surface and banks.

Arquets, les — district: les Hamelins, SPB. Probably from above surname, **Larchier**. Archiers, q.v.

Arrivé, le **Mont** — district: E of les Guelles, SPP. After **Pierre Arrivel**, 1793. mont q.v.

Arsenals — militia depots built in the 19th c. by the States of Guernsey:

1. la **Ville** (l'Hyvreuse) SPP
2. le **Bordage** (Baubigny) SSP
3. les **Beaucamps** C
4. les **Naftiaux** SA
5. les **Islets** SPB

Artur, d'**Anne** — field: la Mazotte, CduV. Surname extinct 17th c. **Artur**.

Artur, d'**Job** — fields: les Rohais, SPP; les Arguilliers, VdelE. Derivation as above.

Assault, l' — fields: le Clos Landais, SSV. 'assault', i.e. misdemeanor.

Assault, l', fontaine — spring: in field at la Flaguée, SSV.

Assecrel, l' — field: le Marais, SSP. Extinct surname **Assequerel**.

Asseline, d' — fields: le Pont Renier, SPP; la Fallaize, SM. Surname extinct 16th c. **Asseline**.

Assiette (f) — site (of building).

Auban, le **Mont** — fields: les Blicqs, les Eperons, SA. Surname extinct 15th c. **Auban**. Also misspelt **Mont Audain**. Earliest ref. **Mont Auban**, 1572. mont q.v.

Aubert, le **Val** — land: les Laurens, T. Extant surname, **Aubert**. val q.v.

Auberts, les — fields: Airport, F. Derivation as above.

Auberts, rue des	road: le Bourg, F.
Aubin, d'	field: le Moulin-de-haut, C. Surname extinct 16th c. **Aubin** (extant Jersey).
Aubrets, les	see **Brayes**, la ruette.
Audain, le **Mont**	see **Auban**, le **Mont**.
Auge, l'	land: les Fontenelles, SSV. 'trough'.
Aumône, l'	fields: E of le Villocq, C; N of les Brehauts, SPB. Presumably rentes owed for alms, or land given in alms.
Aumône, rue sous l'	main road: l'Aumone, C. '**rue de dessous l'Aumône'**. 16th c.
Aumont, la **Cotte**	field: S of le Gron, SSV. Extant surname, **Aumont**. 19th & 20th c. spelling, **Côtes aux Monts**. (after **John Aumont**, SSV, 1516). cotte, q.v.
Aumonts, la **Hougue** des	see Osmonds, les.
Aumonts, les	see **Osmonds**, les.
Aupetit, la **Ville**	see **Ville au Petit**, la.
Autel, l'	megalith: 18th c. expression, literally 'altar'.
Autil, le **Clos**	field: la Marette, SSV. Surname extinct 15th c. **Autil**. clos q.v.
Aval, d'	'lower', all parishes.
Aval, **Maison** d'	house: rue d'Aval, VdelE; N of les Prevosts, SSV; les Messuriers, SPB, former rectory, sold 1441; la Bellée, les Brouards, Pleinmont, T.
Avaleur, le **Long**	cliff: S of la Hougue Anthan, SPB. 'long dipping headland'. Misspelt le **Long Cavaleur**, 19th & 20th centuries.
Avallet, Fief d'	feudal holding of land. Dependency of Fief de Blanchelande, SM.
Aveine, le **Clos** à l'	land: la Hougue au Comte, C. clos q.v.

Avisse, Fief **Dame**	feudal holding of land, le Marais, C: late of **Avisse de Vic**. Now known as Fief **de la Rivière**, C.
Avoine, l'	fields: les Rohais, SPP; les Valettes, le Rignet, SSV; unlocated SA. Surname extinct 15th c. **Aveyne** (**Guillaume Aveyne**).
Axcé, l'	see **Hacsé**.

B

Bachelers, les	fields: le Mont d'Aval, C. Surname extinct 16th c. **Bacheler**(**Guillotin Bacheler 1470**).
Bacheley, la **Hougue**	houses: N of les Rouvets, SSV. Derivation as above. hougue q.v.
Back street	properly '**rue de Derrière'**, SPP.
Baconnerie, la	land: Mill street, SPP. Surname extinct 15th c. **De Bacone**.
Badin, le **Camp**	land: la Claire Mare, SPB. Extinct surname **Badin**. camp q.v.
Bahuchet, de	field: les Landes-du-Marché, VdelE. Origin unknown.
Bailleul, de	fields: les Etibots, SPP/VdelE; S of le Marais, SSP; les Niaux, C; les Clercs, SPB; unlocated SM; la Croix-au-Baillif, Contrée de l'Eglise, SA. Surname extinct 18th c. **Bailleul**.
Bailleul, la **Hougue**	land: S of le Mont Cuet, CduV. Derivation as above. hougue q.v.
Bailleul, le camp	land: les Clercs, SPB.
Bailleul, les **Vaux**	fields: N of les Bémonts, SA. Derivation as above. val q.v.
Bailleul, ruette	land: le Hurel, SM.
Bailleuls, les	district: SA. Derivation as above.

Baillif, au	fields: le Passeur, CduV; Plaisance, les Bruliaux, SPB; Airport, F. Surname extinct 16th c. **Le Baillif.**
Baillif, la **Croix** au	site: wayside cross, SA. Derivation as above. croix q.v.
Bailloterie, la	district: CduV. Extinct surname **Baillot.**
Baincul, sur	land: Jerbourg, SM. Possibly animal pound on top of cliff.
Baisses, ès	fields: Ville ès Pie, CduV; Saints, SM. 'slight hollow'. Recorded ès **Besses**, 1581,1654, 1666.
Baisseurs, les	field: les Bas Courtils, SSV. 'shallow hollow'.
Baissières, les	district: SPP/C/SA. 'low-lying' area. Recorded **Bessières**, 1574.
Balan, de	houses, land: Grange SPP; les Barrats, VdelE; la Quevillette, SM. Surname extinct 16th c. **Ballan** (**John Ballan** 1499).
Balan, la **Houguette**	land: le Port de Nermont, CduV. Recorded 1591. Derivation as above. houguette q.v.
Balan, la **Rocque**	house: l'Ancresse, CduV. Derivation as above. rocque q.v.
Balan, les **Hougues**	land: les Banques, SSP. Derivation as above. hougue q.v.
Ban, au	see **Auban.**
Ban, le Mont au	see **Auban.**
Banque, de	field: le Hamel, VdelE. Surname extinct 17th c. **Banque.**
Banque, la **Haute**	land: rue Godfrey, CduV; la Lague, SPB. 'high' (sand/pebble) bank.
Banques, les	coast, les Banques, SPP; SSP; C; SSV; SPB. Banks of sand, shingle.
Banquet, le	road: Pleinmont, T. 19th c. road name.
Banquette, la	road: Albecq, C. 19th c. road name.

Barbarie, la — house: Saints, SM. 19th c. name for older house.

Barbier, le **Clos** au — house, field: N of la Villette, SM. Surname extinct 15th c. **Le Barbier**. clos q.v.

Barnabey, de — see **Bernabey**.

Barrat, rue du — road: near Albecq, C.

Barrats, les — fields: la Hure Mare, CduV; Sous les Courtils, C. 'drainage ditches' (dykes). see **Dicqs**.

Barrats, rue des — road: now called rue des Barras, VdelE.

Barré, le — fields: les Vardes, SPP; les Marais, SPB. Extinct surname, **Le Barrey**. Recorded, 'qui fut au **Barrey'** 1574.

Barrières de la Ville, les — Marker stones (bornes), SPP. Erected in 1700 to mark the boundaries of the medieval town within which there was no 'préciput', (eldership in inheritance).

Barrières, les — fields: le Mont d'Aval, C. 'enclosure barriers'.

Barsier, fontaine — spring: rue Rocheuse, SPB.

Barsier, le — fields: les Grands Moulins, C; la Rue Rocheuse, SPB. Surname extinct 16th c. **Le Barsier**, also spelt **Barcier** (**Perrotine La Barsière**, 1548).

Barsière, la rue — road: King's Mills, C.

Barthelemie, de — field: S-W of les Villets, F. Surname extinct 14th c. **Bartholomier** (**Colin Bartholomier**, 1331).

Bas Chemin, le — see **Chemin**, le **Bas**.

Bas Courtils, fontaine — spring: les Bas Courtils, SSV.

Bas Courtils, les — see **Courtils**, les **Bas**.

Bas Marais, le — see **Marais**, le **Bas**.

Bas Séjour, le — see **Séjour**, le **Bas**.

Basille, de **James**	field: la Landelle, C. Surname extinct 16th c. **Basille** (**Bazille**).
Basille, la **Mare**	land: la Hougue à la Perre, SPP. mare q.v.
Basses Capelles, les	see **Capelles**, les **Basses**.
Basseterre, la	see **Terre**, la **Basse**.
Bassets, les	fields: les Massies, SSV; La Vrangue, T. Surname extinct 15th c. **Basset**. **Guillebert Basset** 1425, **Pierres Basset** 1535, 1555.
Bassire, la **Mare**	land: unlocated CduV. Surname extinct 16th c. **Bassire**. mare q.v.
Bastien, le	fields: les Reines, F; la Vallée, SM. Forename (m) **Bastien**.
Bastille, la	field: la Villiaze, F. Former house, origin unknown.
Bataille, la	fields: N of Vauvert, SPP; la Grande Rue, SSV; la Claire Mare, SPB; le Vacheul, F; Calais, SM. Surname extinct 16th c. **Bataille** (**John Bataille**, 1536). Battle lane q.v.
Bataille, la ruette	land: between Havilland street & St John street, SPP, Battle Lane q.v.
Bâte, la	fields: Airport, F. Presumably extinct surname. Recorded, les **Bastes**, 1596.
Bâtée, la	fields: la Hougue du Moulin, CduV. Origin unknown. Recorded, la **Bataye**, 1591.
Bâton, de	fields: les Cambrées, SPB/T. Surname extinct 15th c. **Baston**.
Bâton, le **Camp**	land: le Coudré, SPB. Derivation as above. camp q.v.
Bâton, le **Frie**	house: le Frie Bâton, SSV. Derivation as above. frie q.v.
Batteries	coastal batteries, listed by parish:
SPP	Mont Arrivé, Glategny, Hougue à la Perre, le Becquet (Fermain), Fort George (Adolphus, Charlotte, Clarence, Kent, Town Point, York).

SSP	Bellegrève, Delancey, Kempt (la Tonnelle), Spur.
CduV	Beaucette, Chouet (2), l'Ancresse, Half-moon, le Nid de l'Herbe, le Platon.
VdelE	Portinfer.
C	Burton, le Guet, Fort Hommet, Vazon (2).
SSV	le Mont Chinchon, Perelle.
SPB	Brock, l'Erée.
T	les Pezerils, Pleinmont (2), les Tielles.
F	Petit Bot, les Sommeilleuses, la Corbière.
SM	le Bec du Nez, Bon Port, Fermain, Icart (2), Jerbourg (2), Mont Hubert, Moulin Huet (2), Saints (2).
SA	none.

Battle lane — lane: off Grange SPP. 19th c. road name. Translation of surname **Bataille** q.v.

Baubigny, de — district: Baubigny SSP. Surname extinct 14th c. **De Baubigny**.

Baudain, de — field: les Croutes, SPP. Surname extinct 15th c. **Baudain** (**Nicollas Baudein**, 1482). Now extant as **Baudains**.

Baudany, de — field: near la Ville au Roi, SPP. Probably from extinct surname.

Baudu, la **Ville** — district: la Ville Baudu, CduV. Extinct surname **Bodu**, ville q.v.

Baugy, de — fields: N of Sohier, CduV; le Valinguet, C; Jerbourg SM. Extant surname **De Beaugy**.

Baugy, la hure — land: S-W of le Vallon, SM.

Bazin, le **Ménage** — field: le Rocré, C. Surname extinct 19th c. **Bazin**. ménage q.v.

Beaucamp (**s**) les — houses, fields: rue de la Cache, les Vallées, les Beaucamps, district including school and former arsenal, C. Surname extinct 21st c. **De Beaucamp**. Also feudal holding of land, Bordage **Beaucamp**, C. q.v.

Beaucamp, le **Courtil** — house: les Maingys, VdelE.

Beaucettes, les — fields: Beaucette marina, CduV. Origin unknown. Recorded **Biausetez** 1591, **Beausettes**, 1721.

Beaulins, les — house, fields: la Maison de bas, SA. Surname extinct 17th c. **Beauly**.

Beaumont, de — house: near St Anne, C. Early 19th c. house name, 'elegant villa'. Now named la Tiosterie, q.v.

Beaunis, de — field: la Maraive, CduV. Origin unknown. Recorded **Beaunis**, 1591.

Beaupeigne, la fosse de — land: Jerbourg SM. Origin unknown. fosse q.v.

Beaupère, de — fields: le Mont d'Aval, C. Surname extinct 16th c. **Beaupèrre** (**Thomas Biaupèrre** 1470).

Beaurégard, de —
1. house: les Vauxlorens, SPP. 'fine view'. 18th c. house name.
2. la **Tour de Beaurégard**, SPP. Medieval tower in south of town, demolished 18th c.
3. fields: E of le Frie Bâton, SSV. 'good aspect'.

Beau Séjour, de — see **Séjour, Beau**.

Beausouffle, la **Hougue** — land: les Rébouquets, F/SM. Extinct surname **Bausuffle**, 1671. hougue q.v.

Beauvalet, le — land: les Terres, SPP. Surname extinct 12th c. **Bonvalet**.

Beauvalet, le **Moulin** de — district: E of les Padins, SSV. Formerly two watermills now in the reservoir. 'fine valley', or derivation as above. Misspelt as **Beauvallée**, 21st c. Moulin à eau q.v.

Beauvoir, de — house: les Granges, SPP. fields: les Ruettes Brayes, SPP; La Ronde Cheminée, les Longs Camps, SSP; S of le Mont Cuet, la Grande Rue, CduV; rue des Marais, VdelE. Surname extinct 19th c. (**De**) **Beauvoir**.

Beauvoir, Fief de — feudal holding of land, (**Nicollas**) **Beauvoir**, SM.

Beauvoir, la **Hougue** — hillock: W of la Hougue à la Perre, SPP. Derivation as above.

Bec du Nez, le — see **Nez**, le **Bec** du.

Béco, la **Hure** — houses: W of les Rouvets, SSV. hure q.v.

Béco, le — field: les Adams, SPB. Surname extinct 16th c. **Beccot**. Also spelt **Becoq**, **Bequoc**, **Bequot**.

Becquesses, le — land: la Villiaze, SA. Surname extinct 14th c. **Bequisses**.

Becquet, le — fields: Fermains, SPP; le Mont Varouf, rue à l'Or, la Marette, SSV; Le Douit du Moulin, SPB; Les Cornus, le Rocher, SM. Surname extinct 18th c. **Becquet**.

Becquet, le **Moulin** de — watermill: le Douit du Moulin, SPB. Demolished during the German Occupation 1940–45. 'broad flowing stream' or derivation as above. Moulin à eau q.v.

Becqueterie, la —

1. land: former house by watermill (ref. above). SPB.
2. land: former house below les Arquets, SPB. Derivation as above.

Becqueville, de — see **Bequeville**, de.

Becrel, la **Croûte** — land: les Mares Pellées, CduV. Surname ext. 16th c. **Becquerel** (**Guillotin Bequerel**, 1433). Bordage, croûte, q.v. Also feudal holding of land, Bordage **Becquerel**, CduV.

Becville, de — see **Bequeville**, de.

Bégaret, de — field: les Falles, SPB. Origin unknown. Also spelt **Bigarré**.

Bégine, hougue — land: la Hougue du Moulin, CduV.

Begueville, de — fields: la Fallaize, SM. Surname extinct 14th c. **De Begueville**. Also spelt **Becville**,

Becqueville. Also feudal holding of land, Fief **Giffré de Begueville**, SM (now part of Fief de Blanchelande), and feudal holding of land, Fief de **Begueville**, T (now known as Fief des Huit Bouvées).

Béguin, le — former beacons on high ground:

les **Monts**, Delancey Park, SSP
la **Hougue du Moulin**, CduV
la **Hougue Fouque**, SSV (possibly)
à **Pleinmont**, T
le **Mur**, Jerbourg, SM

Béhehe, la hougue — land: le Mont d'Aval, VdelE.

Behier, les hougues — land: Vale Road, SSP.

Béhons, les — land: la Rochelle, CduV. Presumably extinct surname. Recorded les **Bouhons**, 1591.

Béhot, de — field: les Grandes Capelles, SSP. Origin unknown.

Beir, la **Croix** au — site: le Rocher, C. Former wayside cross. Surname extinct 19th c. **Le Beir**. croix q.v.

Beir, la **Croûte** au — fields: W of la Croix au Beir, C. Derivation as above. croûte q.v.

Beir, la rue au — road: former name for **rue des Carterets**, C. Derivation as above. rue q.v.

Belenger, de — fields: la Generotte, C; les Rohais, SA. Surname extinct 14th c. **Belenger**.

Belengère, la **Fosse** — land: les Vallées, C. fosse q.v. Also feudal holding of land, Fief de la **Belengère**, C.

Belette, la — see **Bellette**, la.

Bélial, le — houses, fields: rue du Presbytère, C; le Bélial, SSV. Possibly signifies 'unpleasantness' or immense difficulty in cultivation of heavy land.

Belic — see **Blicq**.

Belle, fontaine — spring: les Maindonnaux, SM.

Belle, le **Grand**
house: la Contrée de l'Eglise, SA. 'large farmyard.'

Belle, le
houses: les Marchez, SPB. 'paved farmyard', alternatively surname extinct 16th c. **Belle**.

Belle, le
house: La Bellée, T. Derivation as above. Also feudal holding of land, Bordage **Pierre Belle**, C.

Bellée, la
houses, land: W of la Forge, T; les Bailleuls, SA. Collective name for houses at le Belle, T.

Bellée, la **Fontaine**
land: bordering road at la Bellée, T. fontaine q.v.

Bellée, le **Vau**
land: les Vauxbelets, SA.

Bellegrève, de
bay: fine sandy beach, SPP/SSP. Also spelt **Belgrave**.

Belles Filles, de
fields: N of la Vrangue, SPP. Possibly rente owed to a female church lay order.

Belles Filles, ruettes des
land: near St Catherine, la Ramée, SPP.

Belles, la rue des
road: 20th c. name for **rue au Feyvre**, C. rue q.v.

Belles, les
house, fields: les Prevosts, SSV. Derivation as above from surname **Belle**.

Belle Voie, la **Croix** de
site: E of ruette du Frocq, C. Former wayside cross, possibly a pilgrimage cross. Also spelt **Belle Voye**. croix q.v.

Bellette, la
fields: le Coudré, SPB; Rue Poudreuse, SM. Origin unknown.

Bellette, ruette à la
lane: unloc. SM.

Bellieuse, la
district: medieval settlement around church, SM. Recorded as **Bellosa** 1267, villate de **Bellousa** 1309. Origin unknown.

Belmont, de
land, road: les Croûtes, SPP. Named after adjacent 18th c. house, **Belmont**, Queen's Road, SPP.

Belval, de — house, fields: la Ville ès Pies, CduV. 'fine valley'. Recorded **Belle Vall**, 1591.

Belvedere, de — house: les Terres, SPP. 'lookout'. 18th c. house.

Bémonts, les — fields: W of les Huriaux, SM/SA. 'fine mount'. Recorded **Beumont**, 1619.

Bémonts, rue des — road: W of les Huriaux, SM. Now called rue des Landes, or Donkey Hill.

Benest, le **Douit** — house, land: le Val, SPB. Derivation from either forename or surname, **Benest**, **Benoist**. Recorded **Benez** 1695, **Benoist** 1699. douit q.v.

Benest, le **Hommet** — tidal off-shore islet: CduV. 'blessed', or from above surname. Hommel, hommet q.v.

Bequepés, **Fief** de — feudal holding of land: le Marais Gouies, SPB. Misspelling of surname extinct 15th c. **Becquisses**, **Becquesses** q.v.

Bequet, le — see **Becquet**.

Bequeterie, la — see **Becqueterie**.

Bequeville, de — field, road: near la Fallaize, SM. Surname **de Begueville**, 1331.

Berger, la **Trappe** — area: Mont Plaisant, C. Also spelt **Bergiers**. Surname extinct 16th c. **Berger** (**Bergier**). trappe q.v.

Bergerots, les — land: E of les Beaucamps, C. Presumably extinct surname.

Bergerots, rue des — road: near les Beaucamps, C.

Bergers, douit des — stream: near le Marais, C.

Bergers, les — fields: N of les Grands Moulins, C.

Bergers, rue des — road: N of King's Mills, C.

Bernabey, de — land: le Pont Renier, SPP; la Ville à Beuf, C. After **Bernabey Le Gersiez** (**Le Jersiais**), 1548.

Bernard, de **Janin**	land: la Ramée, SPP; Victoria Avenue, SSP; les Paysans, SPB. Surname extinct 15th c. **Bernard**, also spelt **Besnard**. saudrée q.v. Also feudal holding of land, Fief **Janin Bernard**, SPB.
Bernard, la **Croix**	site: la Miellette, CudV. Former wayside cross. Derivation as above. croix q.v.
Bernard, la **Hougue**	land: Hacse, CduV. Derivation as above. hougue q.v.
Bernard, le **Clos**	field: Airport, F. 'Derivation as above'. clos q.v.
Bernardière, la	house, fields: les Poidevins, SA. Derivation as above.
Bernauderie, la	house: les Landelles, C. Derivation as above.
Berthe, de	field: le Douit de Moulin, SPB. Origin unknown. Recorded **Berthe**, 1504.
Berthelaye, de	fields: W of les Villets, F. Origin unknown. Recorded **Bertelaye**, 1671. Also spelt **Berthelée**.
Berthelot, la rue	houses, street: SPP. Part of **chemin le Roi**, q.v.
Berthelots, les	field: la Pomare, SPB. Derivation as above
Bertin, courtil	field: la Houguette, SPB. Forename (m), **Bertin**.
Bertozerie, la	houses: les Godaines, SPP. Surname extinct 17th c. **Bertault**. Also spelt **Berthauserie**.
Bertran(d), de	fields: le Hamel, VdelE; les Houmets, C; la Fallaize, Saints, la Fosse, SM. Extant surname **De Bertrand**.
Bertrand, fosse de	land: Jerbourg, SM.
Bertrand, la **Croix**	site: les Camps, SM. Former wayside cross. Derivation as above. croix q.v. Also feudal holding of land, Fief **Bertram**, C.
Bertrand, la **Croûte**	land: la Croix Bertrand, SM.

Besnard, de — see **Bernard**.

Besnard, hougue — land: S of Hacse, CduV.

Besnardière, la — see **Bernardière**.

Besognes, rue des — road: near la ruette du Frocq. C.

Besoignes, les — fields: la Trappe Berger, C. Surname extinct 15th c. **De La Besoigne** (**Jourdian De La Besoigne**, 1470). Also feudal holding of land, Fief des **Besoignes**, C.

Besses, ès — see **Baisses**.

Bessières, les — see **Baissières**.

Bêtes, le **Vaux** — cliff gulley: Jerbourg, SM. 'animal' valley. val q.v.

Bèton, la **Croûte** — land: la Rochelle, CduV. 'concrete', very hard land. croûte q.v.

Bette, douit de la — field: les Maindonnaux, SM.

Bette, la — scrubland: les Talbots, C; le Felcomte, les Monts Hérault, SPB; les Maindonnaux, Icart, SM; les Rohais, SA. Used for hobbled animals. Also extinct surname **De La Bete**.

Bette, rue de la — road: unlocated, SPB.

Beuf, à — fields: les Maindonnaux, SM; les Baissières, SA. Surname extinct 16th c. **Le Boeuf**.

Beuf, la **Rocque** à — house, fields: les Delisles, C. Derivation as above. rocque q.v.

Beuf, la **Ville** à — fields: les Bergers, C. Derivation as above. ville q.v.

Beuf, le **Clos** au — field: la Lande, C. Derivation as above. clos q.v.

Beuf, le **Demi** — fields: la Palloterie, SPB. Fields 'halved'. Derivation as above.

Beuval, de — fields, road: les Sages, SPB/T. Surname extinct 15th c. **De Beuval**.

Beuval, la **Hure**	land: le Campère, SPB. Derivation as above. hure q.v. Also feudal holding of land, Fief de **Beuval**, SPB.
Beuval, le **Douit** de	house, land: les Sages, SPB. Derivation as above. douit q.v.
Beuverie, de	see **Bouverie**.
Beuverie, rue du	road: E of the coast road at Cobo, C.
Beuves, de	field: la Vrangue, F. Surname extinct 15th c. **Beuves** (**John Beuves**, 1425).
Bibert, de	field: les Sauvagées, SSP. Surname extinct 16th c. **Bibert** (**Guillome Bibert**, 1574).
Bichard, de	field: le Ponchez, C. Extant surname **Bichard**.
Bichard, la **Ville**	see **Pichart**, la **Ville**. ville q.v.
Bigard, le	houses: S of la Villiaze, F. 'small enclosed gardens'. Alternatively surname extinct 15th c. **Du Bigard**.
Bigardel, le	garden: le Colombier, T. Diminutive of **Bigard**, q.v.
Bigardiaux, les	gardens: les Reveaux, SPB. 'enclosed steep slopes', la Fosse, SPB. Derivation as above.
Bigoterie, la	house, land: rue Berthelot, SPP. Early 20th c. house name.
Bigotez, la **Croûte** des	land: les Vardes, SSP. Recorded **Collin Bigot** 1478. Surname extinct 15th / 16th centuries. **Bigot**.
Billoterie, la	fields: N of le Bas Marais, SSV. Surname ext. 18th c. **Billot** (**Edmont Billot**, 1628).
Billottes, les	field: les Effards, C.
Binbeurs, les	field: Bon Port, SM. Origin unknown. Possibly the 'shakers', i.e. wind blown land.
Binet, le **Puits**	land: le Bosq, SPP. Surname extinct 15th c. **Binet**.
Birro, de	land: les Arguilliers, VdelE. Origin unknown

Bisson, de	fields: le Houmtel, le Marais, CduV; les Arguilliers, VdelE. Extant surname **Bisson**.
Bissonnerie, la	land: les Ruettes, C; la Pomare, SPB. 'bush, white thorn'. Also spelt **Buissonnerie**, **Buissoniere**. Acbuisson q.v.
Blanc, au	land, fields: la Couture, SPB; le Ponchez, C. Extant surname **Le Blancq**.
Blanc Musel, le	see **Musel**, le **Blanc**.
Blanche(s) **Carriere**(s), les	see **Carrierre**, la **Blanche**.
Blanche(s) **Rocque**(s), les	see **Rocques**, les **Blanches**.
Blanchelande, de	feudal holding of land, Fief de **Blanchelande**, SM. Formerly held by Abbot of Blanchelande, Normandy.
Blanches Pierres, les	see **Pierres**, les **Blanches**.
Blanches Rocques, la **Croûte** des	land: below le Câtillon. SPB.
Blanches Terres, les	see **Terres**, les **Blanches**.
Blanches, les	district: les Blanches, SM. Recorded, properties owned by Blanche family 1511–1913. Surname extinct 20th c. **Blanche**.
Blancq, la **Croûte** au	land: Chemin le Roi, F.
Blancs Bois, les	house: rue Cohu, C. 'white woods'. Late 18th c. name.
Blanloteries, les	field: N-E of Saumarez Park, C. Surname extinct 17th c. **Blanlot (Jeremie Blanlot**, 1639).
Blête, la	fields: la Devise, SPB; S of les Laurens, T. 'turved'.
Blicq, de	fields: les Rouvets, SSV; le Russel, F. Extant surname **Blicq**, spelt **Belic** until 18th c.
Blicq, le **Mont**	fields: les Nicolles, SM. Derivation as above. mont q.v.

Blicqs, les	houses, fields, road: les Blicqs, SA. Derivation as above. Vauxbelets q.v.
Bloche, la	field: le Chene, F. Origin unknown.
Blondel, de	fields: Bordeaux, CduV; les Touillets, la Porte, C; les Frances, SSV; la Pomare, la Dévise, SPB; rue des Pointes SA. Extant surname, **Blondel**. Also feudal holdings of land, Fief des **Dix Quartiers Blondel**, SSV, and Fief de **Thomas Blondel**, SPB/T.
Blondel, la **Croûte**	land: le Grand Douit, SSV.
Blondiaux, les	fields: Croûte Becrel, CduV; les Grands Moulins, C; les Vauxbelets, SA. Plural form of **Blondel**. Also spelt **Blondios**.
Bô, le **Moulin** de **Petit**	formerly two water mills, upper and lower. Moulin à eau q.v.
Bô, **Petit**	district: Petit Bot, F/SM. 'small woodland'.
Bocage, le	houses: les Marchez, SPB; la Fosse, SM.
Boe, la	see **Boue**, la.
Boemy, les **Camps**	land: Pleinmont, T. From extinct surname **Boemy**. camp q.v.
Boeuf, hougue au	land: unlocated SA.
Bois, la **Croix** au	site: E of la Maraitaine, CduV. Former wayside cross. croix q.v.
Bois, la **Hougue** du	land: St Clair, SSP. Surname extinct 16th c. **Bois**, **Du Bois**.
Bois, les **Blancs**	see **Blancs Bois**, les.
Boisselée, la	house: Saints, SM. Owing feudal due of 1 bushel (boissel, boisseau) of wheat rente per annum on Vingt Boisselée, part of Fief de Blanchelande.
Bon Air, de	houses: Queen's Road, SPP; les Adams, SPB; Sausmarez, SM. 'salubrious'. 19th c. house name.

Bon Enfant, de	field: la Rochelle, CduV. Extinct surname **Bonenfant**.
Bon Port, de	headland: E of Saints, SM. Overlooking anchorage.
Bon Répos, de	haven: la Corbière, F. 'good rest' (ironic).
Bonamy, de	house, fields: Ashbrook, SSP; la Pomare, SPB; le Russel, F; la Villiaze, SA. Surname extinct 19th c. **Bonamy**.
Bonne Femme, de	see **Femme**, de **Bonne**.
Bordage(**s**) le(s)	land on feudal holding held by bordier of Fief as a perquisite for issuing summonses etc., all parishes. The plural form Bordages is often used for several fields of that name in the same area, and usually connected with a Bordage in the vicinity.
SPP	Bordage **Cornet**, Bordage **Durand**, Bordage **Landry**, Bordage **Lesant**, Bordage **Rungefer**, Bordage **Trousse**.
SSP	Bordage **Fantôme**, Bordage **Geffrey** (Guiffre), Bordage (Jourdain)**Testart**.
CduV	Bordage **Becrel** (Becquerel), Bordage **Rebours**, Bordage **Etienne Renost** (Renault), Bordage **Sallemon**, Bordage **Giffart**, Bordage **Ricart Nant**, Bordage **Galliot**.
VdelE	None.
C	**Chef** Bordage (of Fief le Comte), Bordage **Allez**, Bordage **Beaucamp**, Bordage **de Nordest**, Bordage **Pierre Belle**.
SSV	Bordage **Jourdain David** (Davy), also one unnamed.
SPB	Bordage **De l'Erée**, also one unnamed.
T	Bordage **d'Emon De La Rue**.
F	Bordage **Troussey**, Bordage **Videclin**.
SM	Bordage **au Bouteillier**, Bordage **Saete**.
SA	Bordage **Almenac** (Almena).

Bordage **Brisepik**, Bordage **Pantalaron** (or **Vauquiedor**) SPP are unlocated.

Bordage(s) – property associated with that name:

houses, land: le Bordage, SPP	Bordage **Cornet**
house, land: la Ramée, SPP	(Bordage unidentified)
houses, arsenal, land: Baubigny, SSP	Bordage **Geffrey**
houses, land: les Bordages, SSV	(unnamed Bordage)
houses, land: le Bordage, SPB	(unnamed Bordage)
fields: l'Alleur, T.	Bordage **d'Emon De La Rue**
part of 2 fields: les Villets, F	Bordage **Videclin**
house, land: la Croisée, F.	Bordage **Troussey**
house: le Bordage (la Bellieuse), SM	Bordage **au Bouteillier**
field: les Mézières, SA	(Bordage unidentified)

Bordage, rue du — road: W of Fountain street, and road now called First Tower Lane, SPP; N of la route du Caudré, SPB; at les Villets, F; la Bellieuse, SM.

Bordeaux, de — houses, land: Bordeaux, CduV; rue Cohu, la Lande, C; le Grand Belle, SA. 'minor border lands'. Plural of **Bordel**, q.v.

Bordeaux, hougue de — land: Bordeaux, CduV.

Bordel, le — fields: la Ville au Petit, les Etibots, SPP; le Bordel, CduV; les Nicolles, SM. Diminutive of **Borde**, see **Bordeaux**.

Bordes, les — houses, land: N-W of le Catioroc, SSV. 'unenclosed border lands' on coast.

Bordes, rue des — road: N-W of le Catioroc, SSV.

Bos, les **Longs** — fields: N of la Hougue Fouque, SSV. 'long, tall woods'. Recorded les **Longs Booz**, 1586. Misspelt **Longbeaux**, 20th c.

Bos, rue des **Longs** — road: N of la Hougue Fouque, SSV.

Bosq, le — houses, land: E of les Canichers, Collings Road, SPP; rue Piette, C. Surname extinct 15th c. **Du Bosc**.

Bosq, le **Grand** — house: N of St Julian's Avenue, SPP. Central building in former Royal Hotel. Derivation as above.

Bot, de **Petit** — see **Petit Bot**, de.

Boudin, le **Douit**	field: S of les Houmets, C. Surname extinct 17th c. **Boudin**. douit q.v.
Boudin, rue du **Douit** de	road: S of les Houmets, C.
Boue, la	1. rock: low lying offshore rock (covered in weed) exposed at low water. 2. houses, land: W of les Vinaires, SPB. 'low lying swamp' (in valley). Recorded **Boe**, 1671.
Boue, rue de la	road: W of les Vinaires, SPB.
Bouel, le **Long**	see **Bourel**, le **Long**.
Bouet, le	houses, land, road: le Bouet, SPP. Surname extinct 17th c. **Du Bouet (Simon Bouet**, 1610).
Bougourd, de	field: la Bailloterie, CduV. Extant surname **Bougourd**.
Bouguenôt, la	land: Jerbourg, SM. Origin unknown.
Bouilleuse, la **Croûte**	fields: les Grands Capelles, SSP. 'bubbling'. croûte, q.v.
Bouillon, le	houses, fields: Longstore, la Ville au Petit, SPP; les Osmonds, St Clair, SSP; le Villocq, les Mourains, la Lande, les Delisles, C; les Jenemies, SSV; les Adams, SPB; le Bouillon, SA. 'spring'. fonderille, fontaine q.v.
Bouillonne, les **Dégrais** de la	steps: les Petites Fontaines, SPP. la **Vallée** du **Mont Durand** q.v.
Bouillonne, ruette de la	land: Mont Durand, SPP.
Bouillonnière, la	fields: la Ville au Roi, SPP; le Vaugrat, SSP; les Delisles, C; la Rousse Mare, SPB; les Laurens, T; Farras, F. 'area of springs'.
Bouillonnière, rue du	road: unlocated, F.
Boulains, fontaine	spring: W of Fauquet Valley, C.
Boulains, les	fields: S of les Grantez, C. Surname extinct 16th c. **Boulain**.

Boulevards, les	embankments: at bays in all parishes, protection against coastal erosion. Also 18th c. defensive works.
Boullerie, la	land: le Four Cabot, SA. Surname extinct 16th c. **Boulez**.
Boullet, le **Bas**	land: le Hamel, VdelE. Derivation as above.
Boulliers, le **Camp** ès	land: la Rocque, SPB. Surname extinct 15th c. **Boullier**, (extant Jersey).
Bourdeaux, les **Vaux**	field: le Grand Belle, SA. See **Bordeaux**.
Bourdon, le **Pré**	field: le Frie Baton, SSV. Surname extinct 17th c. **Bourdon**.
Bourel, le **Long**	field: Paradis, CduV. Surname extinct 16th c. **Bourel**.
Bourg, le **Val** au	slopes: Saints, SM.
Bourg, le	houses, land: le Bourg, F. 'central settlement in reclaimed land'.
Bourg, les **Fontaines** au	land: les Vallées, C. fontaine, val q.v.
Bourgaize, de **Jean**	field: les Falles, SPB. From surname **Bourgeois**, now spelt **Bourgaize**.
Bourgeois, au	fields: la Hougue Anthan, SPB; les Sages, T; la Villiaze, F. Derivation as above.
Bourgeron, la **Croûte**	field: les Simons, T. Surname extinct 16th c. **Bourgeon**. croûte q.v.
Bourgs, les	houses, land: les Bourgs, SA. Surname extinct 17th c. **Le Bourg**.
Bournel, Fief de	feudal holding of land. Otherwise known as Fief de **Bruniaux de Nermont**, q.v.
Bourse, la	see **Fonds** de la **Bourse**, la.
Bout, le	house: le Grée, T. At 'end' of parish of Torteval.
Boutefèves, les	fields: les Blanches Pierres, SM. Possibly 'incendiary'. Recorded de **Boutte Fayve**, 1546.

Bouteillier, le **Bordage**	feudal holding of land, unlocated SM. Surname extinct 16th c. **Le Bouteillier**. Now spelt Le Boutillier. Bordage q.v.
Bouteillier, le **Clos** au	house: le Bordage (la Bellieuse), SM. Derivation as above. Bordage au **Bouteillier** q.v.
Bouteilliers, les	fields: les Cordiers, SPP; la Vieille Rue, SSV. Derivation as above.
Bouvée, fontaine	spring: E of la Bouvée, SM.
Bouvée, la	1. superficial measurement of land: 20 vergées. 2. houses, fields: le Pont Perrin, VdelE; le Neuf Chemin, les Bordages, la Bouvée, la Grande Rue, SSV; Pleinmont, T; la Bouvée, SM. fontaine q.v 3. feudal holding of land, Bouvée **Marquand**, SSV. 4. feudal holding of land, Fief de la **Bouvée Duquemin**, SPB, formerly part of Fief au **Canely**. 5. feudal holding of land, Fief des **Huit Bouvées**, SPB, associated with Fief St Michel. 6. feudal holding of land, Fief des **Huit Bouvées**, T, formerly Fief de **Begueville**. Begueville, Canely, Duquemin, Marquand q.v.
Bouverie, de	land: rue de Bouverie, C. Surname extinct 16th c. **Beverie** (**Thomas Beverie**, 1551).
Brache, de	land: les Fries, C. Earliest ref. 1833. Extant surname **Brache**.
Bracherie, la	house (demolished): les Annevilles, SSP. Ref. 1686. Derivation as above.
Brasserie, la	house (demolished): le Bordage, SPP. 'brewhouse'.
Braye, camp du	land: l'Erée, SPB.
Braye, la route du	road: le Braye du Valle, CduV. Formerly rue **Mollet** in W, rue **Allez** in E.

Braye, le	1. straits: le **Braye du Valle**, formerly separating le Clos du Valle from Guernsey, within which, after reclamation, roads built – rue **Allez**, route **Carré**, rue **Mollet** (after the contractors), route **Doyle** (now route **Militaire**). Allez, Carré, Doyle, Mollet, route q.v. 2. straits: le **Braye de Lihou**, separating Lihou from Guernsey. 3. fields: la Grosse Hougue, SSP; les Mares Pellées, les Monmains, CduV; l'Erée, SPB. Adjacent to the straits (le **Braye**).
Brayes, la ruette	road, land: les Ruettes Brayes, SPP. Surname extinct 15th c. **Bret**, **Aubret**.
Brayes, les	fields: les Ruettes Brayes, SPP/SM.
Brazart, le **Camp**	land: les Blancs Bois, C. Surname extinct 15th c. **Brassart**.
Brazar[t], le **Douit**	stream: les Blancs Bois towards la Charruée, C/VdelE. Derivation as above. douit q.v.
Brèche, la	'opening to field', all parishes. Uncultivated. Also spelt **Brècque**, **Brèque**.
Brécley, le	fields: les Maindonnaux, SM. Diminutive of **Brèche**, q.v. Recorded **Bréquelet**, 1669, **le Brécley**. Bréquet q.v.
Brehaut, de **Collas**	fields: la Mare, SPB; la Corbière, F. Extant surname **Brehaut**.
Brehaut, de **Jean**	field: les Marchez, SPB. Derivation as above.
Brehaut, de **Julien**	field: les Massies, SSV. Derivation as above.
Brehaut, de **Thomas**	field: Calais, SM. Derivation as above.
Brehaut, la **Croix**	site: E of les Brehauts, SPB. Former wayside cross. Derivation as above. croix q.v.
Brehaut, la **Fosse**	field: Paradis, Cdu V. Derivation as above. fosse q.v.

Brehaut, le **Camp**	field: rue Sauvage, SSP. Derivation as above. camp q.v.
Brehaut, le **Camp Massy**	field: le Haut Chemin, SPB. After **Massy Brehaut** (no date). Massy q.v.
Brehaut, rue de	road: S-W of route de Plaisance, SPB.
Brehauts, les	houses, fields: le Douit, VdelE; N of Saumarez Park, C; les Brehauts, SPB; la Croisée, T; la Corbière, F; le Hurel, SM; les Rohais, SA. Extant surname **Brehaut**.
Brèquet, le	field: N of Saumarez Park, C. Diminutive of **Brèche**.
Bretagne, de	fields: les Grandmaisons, SSP; les Caches, SM. Presumably from extinct surname (**De Bretagne**).
Breteures, les	field: Icart, SM. Presumably from extinct surname.
Breton, de	fields: les Duvaux, les Osmonds, SSP; Baugy, CduV; les Nicolles, SM; le Camp Tréhard, les Vauxbelets, les Bourgs, SA. Extant surname **Breton** (**Le Breton**).
Breton, de **Collas**	field: les Houards, F. Derivation as above.
Breton, de **Colliche**	land: les Etibots, VdelE . Derivation as above. Also feudal holding of land, Fief au **Breton**, C.
Bretonnerie, la	house, fields: les Pelleys, C. E of les Naftiaux, SA. Derivation as above;
Brets, au	fields, road: ruette Brayes, SPP; S of la Fosse, SM. Surname extinct 15th c. **Le Bret** (**Michel Le Bret**, 1517) Also feudal holding of land, Fief au **Bret**, SM.
Brigade, la	house: S of la Croix au Bailiff, SA. 18th c. house name.
Brilliant, de	see **Bryant**.
Briocq, **St.**	house, land, fields: S of les Clercs, SPB. After adjacent Chapelle de St Briocq. Chapelle q.v.

Briquerie, la	land: Fort road, le Bouet, SPP; S of les Maumarquis, SA. 'brickworks'. Also spelt **Briqueterie**.
Briquet, le	house, land: les Prevosts, SSV. Origin unknown. Earliest ref. 1797.
Bris, de	fields: la Couture, SPP; les Quartiers, SSP. Surname extinct 17th c. **Bris** (**Guillaume Bris**, 1594).
Brisepik, le **Bordage**	unlocated, SPP. Connected with Bordage **Vauquiedor**, 1574. Bordage q.v.
Brock battery	battery: Rocquaine coast road. SPB. 18th c. Rebuilt in 1950s as a sea defence.
Brock road	roads: les Gravées, SPP; New road, SSP. Named after Major-General Sir Isaac Brock and his brother Daniel de Lisle Brock, a 19th c. Baillif.
Brockhurst	house: Grange, SPP. 18th c. barn conversion, named after owner, William Brock. Extant surname **Brock**.
Broderie, la	field: la Claire Mare, SPB. Origin unknown.
Brouard(s), les	house, fields: les Sarreries, SPB; l'Alleur, T; les Merriennes, les Houards, Airport, F; la Villiaze, SA. Extant surname **Brouard**.
Brown, la rue	road: near St James's street, SPP, now lost.
Bruèlles, les	fields: la Colombelle, SM. Origin unknown.
Brûliaux, les	houses, land: E of la Palloterie, SPB.
Brûliaux, sur les	land: N of la Prévôté, SPB. Cliff land cleared by burning during the middle ages.
Brûlin, la **Vallée** au	field: les Vallées, C. Presumably extinct surname **Brullin**. vallée q.v.
Brun, au	field: les Monts, Delancey, SSP. Extant surname **Le Brun**.
Brun, la **Fontaine** au	field: la Rivière, C. Derivation as above.
Brun, le **Clos** au	field: le Marais, C. Derivation as above.

Bruniaux de **Nermont**, Fief de	feudal holding of land. VdelE.
Bruniaux, Fief de	feudal holding of land. SM.
Bruniaux, la **Croûte** des	land: le Marais, Houmet Quenot, VdelE.
Bryant, de	field: les Martins, SSP. Surname extinct 18th c. **Bryant**. Misspelt as **Brilliant**, 19th c.
Buffardière, la	see **Busardière**, la.
Buis Rocques, les	see **Rocques**, les **Buis**.
Buisson, camp du	see **Bissonerie**, la.
Buissonnière, la	see **Bissonnière**, la.
Bulaban, le **Marais** de	fields: S of les Grandes Mielles, VdelE. Surname extinct 14th c. **Bulaban**. marais q.v.
Bullaban, douit de la	field: les Arguilliers, VdelE.
Bulwer avenue	road: E of la Longue Hougue, SSP. Early 20th c. road name, after Lt. Gen. Sir E Bulwer, KCB, Lt. Gov. of Guernsey 1889–1894.
Bunel, de	field: N of rue Cauchez, SM. Surname extinct 15th c. **Bunel**.
Bunel, la rue	road: E of les Vardes, SPP. Derivation as above.
Buquettes, les	fields: N of Fermains, SPP. Origin unknown.
Burgel, de	fields: le Douit du Moulin, SPB. Origin unknown.
Burnt lane	1. roadway: Vauvert, SPP. Properly 'rue des maisons brulées' (road of burnt houses). 2. road: la Longue Rue, SM, after a 1930s shed fire.
Buroque, de	see **Rocques**, les **Buis**.
Busardière, la	fields: le Neuf Chemin, SSV; W of les Buttes, T. Surname extinct 16th c. **Busard**.

Buttes au **Cauf**, les	see **Cauf**, les **Buttes** au.
Buttes, les	houses, fields: butts for parish archery practice. la Butte SPP les Capelles SSP la Hougue du Moulin CduV rue de la Cache C les Buttes SSV les Buttes SPB les Buttes T les Landes F les Mouilpieds SM rue Frairie SA.
Buttes, les	road: W of Torteval Church, T.
Buttière de terre, la	land: all parishes. 'headland' at end of **camp de terre** (strip of land). courtil q.v.

C

Cabart, la **Planque**	field: le Douit, SSV. Surname extinct 15th c. **Cabart**, also spelt **Quabart**. Planque q.v.
Caboche, la	field: la Turquie, CduV. 'fertile knoll'. First recorded 1755.
Cabot, le **Four**	area: le Four Cabot, SA. Possibly outdoor 'oven' connected with surname extinct 16th c. in Guernsey. **Cabot le Roi**, SA. 1610. Or corruption of **carrefour Cabot**.
Cachart, de	field: S of les Maumarquis, SA. Surname extinct 17th c. **Cachart**.
Cache Vassal, la	see **Vassal**, la **Chasse**.
Cache Vassal, la fontaine	spring: St James Street, SPP.
Cache, rue de la	road: E of Castel Church, C.
Cache, une	houses, fields, all parishes, used for access and exit, i.e. a permanent track with banks

	or hedges on both sides giving access and exit to land-locked property, privately owned. chasse, q.v.
Cachebrée	see **Cauchebrée**.
Caches, les	houses: two noteworthy houses in SM, named in 17th c., being at access and exit to two 'caches'. Jean Bonamy, owner of one, made a pilgrimage to the Holy Land in 1500. Also houses N of Les Prevosts, on the Castel/St Saviour's boundary and at les Brehauts, SPB.
Caches, les	houses, fields, all parishes. Access and exit by two or more caches.
Caen, Fief de **l'Abbesse de**	see **Abbesse**.
Cailles, les	field, road: S of les Effards, C. Surname extinct 15th c. **Chaille**. Also spelt **Chailles**.
Caillet, de	field: North side, CduV. Surname extinct 17th c. **Caillet**.
Cailleterie, la	former house, fields: rue des Cottes, SSP; rue du Val, SPB. Derivation as above.
Cainette, la	field: Bordeaux, CduV. Possibly 'rock'. Recorded **Quainnette**, 1591.
Calais, de	houses, fields, road: Calais, SM. Surname extinct 20th c. **De Calais**.
Calistre Lihou, de	see **Lihou**.
Cambrai	house: le Bigard, F. Renamed after First World War.
Cambrées, les	houses, fields: SPB/T. Surname extinct 16th c. **Cambré**.
Cambrettes, hougue des	land: unlocated SA.
Cambrettes, les	fields: SA. Diminutive of **Cambré**. Also feudal holding of land, Bouvée des **Cambrés, T**.

Cambridge Berth — harbour: SPP. Late 19th c. Named after Duke of Cambridge.

Cambridge Park — park: SPP. Derivation as above.

Cammu, rue au — fields, road: S of rue des Grons, SM. Surname extinct 16th c. **Camus**. rue q.v.

Camp au Prêtre, rue du — road: S of les Arquets, SPB.

Camp Bailleul, rue du — road: unlocated, SPB.

Camp d'Ebat, rue du — road: E of les Fontaines, T.

Camp de **Haies**, Fief des — feudal holding of land. Dependency of Fief de Rozel, VdelE.

Camp de terre (m) — 'strip' of unenclosed land in riage (q.v.) within a campagne (large district) or territoire (area of land).

Camp Dry, ruette de — land: les Landes, CduV.

Camp du Roi, le — houses, gardens: strip of land N of road junction, le Camp du Roi, VdelE. Reputed to have been the medieval market centre, les Landes du Marché. Sold by the Crown, mid-19th c.

Camp Henry, rue du — road: unlocated, T.

Camp Massy Brehaut, rue du — road: W of the route des Paysans, SPB.

Camp Rougier, la croute du — land: near Mont Crevel, CduV.

Campagne de **Pleinmont**, la — fields: Pleinmont, T.

Campagne des **Hougues**, la — fields: les Hougues, C.

Campagne, la — fields: les Hougues, C; le Felcomte, la Hougue Anthan, SPB; Pleinmont, T; les Camps, SM. 'plain, flat fields'. Originally a large area of unenclosed land.

Campère, le	house, field: N of les Clercs, SPB. Earliest reference 1727. Origin unknown.
Campere, rue du	road: E of les Clercs, SPB.
Camposo, le	see **Oso**, le camp.
Campozon, le	see **Oso**, le camp.
Camprots, les	see **Rots**, les camps.
Camps, **les**	fields, all parishes. Also feudal holding of land, Fief formerly held by **Richard des Camps**, now a dependency of Fief de Blanchelande, SM.
Camps, la route des	road: S of la Grande Rue, SM.
Camps Collette Nicolle, les	road: E of Collings Road, SPP.
Camps Collette Nicolle, rue	road: N of le Rohais, SPP.
Camps du Moulin (à vent) les	see **Moulin** (à vent) les **Camps** du.
Camps, les **Cours**	land: la Hougue, le Marais, C; le Gron, SSV; les Brehauts, SPB. 'short strips'. Misspelling of courtes, meaning short.
Camps, les **Grands**	land: Chouet, CduV; les Mares, SPB; Rougeval, T. 'large strips'.
Camps, les **Longs**	land: N of la Vrangue, St Catherine, la Pitronnerie, SPP; la Longue rue, les Capelles, SSP; les Vallettes, les Rouvets, SSV; le Caudré, SPB; les Houards, F; le chemin le Roi, le Coin Fallaize, SM; les Fauconnaires, SA. 'long strips'.
Camps (**Massy**), les	see **Brehaut**, le **Camp Massy**.
Camps, les **Plats**	land: les Comtes, SSV; la Houguette, SPB. 'flat strips'.
Camps, les **Profonds**	land: la Fosse, SM. 'steep strips'.
Camps, les **Ronds**	land: les Grands Moulins, C. 'rounded strips'.

Camps Rulées, la hure des	land: les Laurens, T.
Camps, les **Tors**	land: le Dehus, CduV; les Houmets, C; le Coin Fallaize, SM. 'twisted strips'.
Camps traversains, les	'crossings' or way over 'strips' of land: SPP.

Camps (de terre) with proper names listed by parish, see also individual entries;

SPP	Camp **de la Houguette**, Camps **de la Mer**, Camps **Collette Nicolle**, Camps **Padart**, Camps **Pendarts** (**Padart**), Camp **à la Longue Raye**.
SSP	Camp **Brehaut**, Camp **Code** (**Gode**), Camp des **Croquets**, Camp **Jouanne**, Camp **au Mière**, Camp **de Noirmont** (**de Nermont**), Camp **Queripel**.
CduV	Camp **Clos**, Camp **Dolent**, Camp **Dry**, Camp **Giffré**, Camps **Morvels**, Camp **Ori**, Camp**oso** (**Camp Posso**), Camp **au Passeur**, Camp **Rouget** (**Roûgier**).
VdelE	Camp **Brazar**(**t**), Camp des **Haies**, Camp au **Jersiez**, Camp du **Roi**, Camp **Rouf**.
C	Camp **du Buisson**, Camp **des Corvées**, Camp **de la Cure**, Camps **du Gripé**, Camp **Heulle**, Camp **de la Hougue ès Dames**, Camp **de la Hougue Collin Denis**, Camp **de la Houguette au Tellier**, Camp **des Jaonnets**, Camp **de la Rocque Gohier**, Camp **Ruley**, Camp **du Serqueur** (**Cerqueur**), Camp **au Vesiez**.
SSV	Camp **Clanque**, Camp **Corbin**, Camp **du Douit**, Camp **de Hue**, Camp **du Mont le Val**, Camps **au Nort** (**au Nô**), Camp**ozon**, Camp **Paté**, Camps **Sampson**, Camps **du Val au Colomb**, Camp **Yvêlin**.
SPB	Camp **Badin**, Camp **Bailleul**, Camp **du Braye**, Camp **du Cocq**, Camp **Corbin**, Camps (**du fief de**) **Suart**, Camp **de la Forfaiture**, Camp **de Forge**, Camps **des Huriaux**, Camp **Massy** (**Brehaut**), Camp au **Prêtre**, Camps **Rougier**, Camps **du Trépied** (**au fief le Comte**), Camp **Vallen**, Camp **Varouf**.

T	Camps **Boemy**, Camp **d'Ebat**, Camp **Henry**, Camp **au Pelley**, Camp **du Portelet**, Camp **Rocqueux**, Camps **Rulées**.
F	Champ**huet** (Camp Hue), Camps **Lucas**, Camps **Marchants**, Camp **à la Pellée**, Comp**rocqueux** (Camp Rocqueux), Camp**rots** (Camp Rots).
SM	Camp **Croix**, Camp **du Douvre**, Camp **des Fées**, Camp **Follet**, Camp **Fournie**, Camp **des Frênes**, Camp **du Gal**, Camps **du Grand Chemin**, Camps **à la Grange**, Camps **sur la Haie**, Camp **au Houguillon**, Camp **en Icart**, Camp **à Jaon**, Camps **à Labour**, Camps **du Moulin (à vent)**, Camps **ès Pointes**, Camps **au Sommet de la Petite Porte**, Camp **Putron**, Camp **au Saunier**, Camp **de la Trappe**, **Campauvère** (Camp au Vère).
SA	Camp **au Chêne**, Camp **Joinet** (Camp Jehannet), Camp **à Porc**, Camp **Tréhard**.
Camptréhard, le	see **Tréhard**, le camp.
Camu, la hure au	see **Cammu**.
Candie, de	houses: Candie, SPP; Candie road, C/SA. Earliest reference 1686.
Canicher, la hougue de	land: les Canichers, SPP. Surname extinct 16th c. **Le Cannicher**.
Canichers, les	houses, land: S of les Amballes, SPP. Derivation as above.
Canivet, de	land: Fort George, SPP. Surname extinct 19th c. **Canivet**.
Cannevière, la	land: le Tertre, CduV; la Planque, C; Calais, SM. Also spelt **Canvière**. Medieval hemp or flax area (modern French chenevière). Also feudal holding, Fief de la **Canvière**, C.
Cannevière, rue du	road: unlocated, C.
Canniveaux, les	field: les Poidevins, SA. Derivation probably as above.

Cantereine, de — houses, fields: rue des Marottes, C. Recorded 1470, 1634. Also house, fields, water mill; N of les Sages, SPB. See Quanteraine.

Canurie, la — house, land: la Croûte Becrel, CduV. Surname extinct 15th c. **Le Canu**.

Canus, les — house, fields: les Canus, SSP; la Devise, SPB. Derivation as above.

Capelle Dom Hue, — see **Dan Hue**.

Capelles, les — district, fields: les Capelles, SSP; la Mazotte, CudV; la Villiaze, SA. Surname extinct 17th c. **Capelle**. Thought in later 18th c. to be derived from 'chapelle', and hence the feminine form in all usages, still used today.

Capelles, les **Basses** — area: N of les Capelles, SSP. 'bottom'.

Capelles, les **Grands** — area: E of les Hautes Capelles, SSP. 'big'.

Capelles, les **Hautes** — area: les Hautes Capelles schools, SSP. 'top'.

Capelles, les **Petites** — area: N of les Hautes Capelles, SSP. 'small'. The above descriptions originally related to fields.

Capis, Fief de — feudal holding of land. Dependency of Fief de Blanchelande, SM.

Caquière, la — see Quatière, la.

Carantaine, la — former fields: Croix Bertrand, SM. Also spelt **Quarantaine**. '40 eggs' (rente).

Cardeau, le — fields: S of les Villets, F. Origin unknown.

Cardin, de — field: les Sauvagées, SSP. Forename **Cardin** (**Cardin Fautrart**, 16th c.).

Cardronnet, le — land, fields: l'Hyvreuse, SPP; Paradis, CduV; les Bruliaux, SPB. Surname extinct 15th c. **Cardonnel**. Recorded **Cardonnet** 1574, 1591, 1685. Spelt **Cardronnet**, **Chardronnet**, 19th & 20th centuries.

Carduval, le	field: les Landes, CduV. 'quarter slope'. Recorded du **cart du Val**, 1654.
Careening hard	dock: SPP harbour.
Caretiers, les	see **Quartiers**, les.
Carette, la	see **Querette**, la.
Carey(e), de	land, fields: les Croûtes, SPP; la Gallie, SPB; les Ruettes, les Baissières, SA. Extant surname **Carey (e)**. Always spelt Careye or Careie before mid-17th c.
Caritey, de	see **Queritez**, les.
Carmel, de	buildings: Carmel chapel, (now renamed Zion), SSV; rue Cauchez, SM. 19th c. name.
Carpentier	see **Querpentier**, le.
Carpentier, Fief au	feudal holding of land. Also known as Fief au **Quarpentier**. Dependency of Fief de Rozel, VdelE.
Carré, la route	road: SSP. Named after one of the contractors who reclaimed the Braye du Valle c. 1806.
Carrée Pièce, la	see **Pièce**.
Carrefour, le	houses, land: harbour, SSP; les Goddards, C; les Prevosts, SSV; Ste Hélène, les Bailleuls, SA. 'crossroads'.
Carrefour au **Lièvre**	junction: between Fort road and route de Sausmarez, SPP/SM boundary. Derivation as above.
Carrefour des **Trois Vues (Voyes)**, le	see **Vues**, les **Trois**.
Carrefour, le **Grand**	junction: between rue des Forges (Smith street), la Grande Rue (High street) and le Pollet, SPP. 'cross roads'.
Carrefour, le **Petit**	junction: between le Bordage, rue de la Fontaine (Fountain street), rue du Marché (Market street) and le Haut Pavé (Mill street), SPP. Derivation as above.

Carrel, le	land: les Basses Capelles, SPP; l'Erée, SPB; Pleinmont, T. 'square land'.
Carriaux, les	fields: Saumarez park, les Pelleys, les Grantez, les Hougues, C; la Mare, les Prevosts, le Haut, la Marette, SSV; les Vauxbelets, SA. 'shallow gravel pit'.
Carrière(**s**), la (les)	land, all parishes. 'shallow quarry (ies)', charrière q.v.
Carrière, la **Blanche**	see **Charrière**, la **Blanche**.
Carriot, la rue	see **Chariot**, la.
Carteret, la **Mare** de	land: la Mare de Carteret, C. former pond. mare q.v. Extant surname **De Carteret**.
Carterets, la rue des	road: les Carterets, C. Formerly rue au **Beir** q.v. Derivation as above. Also feudal holding of land, Fief **de Carteret**, C.
Carterets, les	land: S of la Mare de Carteret, C. Derivation as above.
Carterie, la	see **Charterie**, la.
Cartiers, les	see **Quartiers**, les.
Caruée, la	see **Charruée**, la.
Carupel, **de**	see **Queripel**.
Cas Rouge, le	house, land: le Bigard, F. Possible corruption of 'carrefour', meaning crossing place. Earliest reference 17th c.
Cas, la **Rocque** ès	land: l'Ancresse, CduV. Surname extinct 15th c. **Cas** (**Quas**). Recorded **Quas** 1591, **Juas**. hougue q.v.
Casernes, les	houses, land: l'Erée, Rocquaine, SPB. Former 18th c. barracks, now both renamed.
Cassot, de	land: New road, SSP.
Cassot, le **Coin**	land: les Nicolles, SSP. Surname extinct 16th c. **Cassot**. coin q.v.

Castel Jean Hue, le	land: les Monts Herault, SPB. 'earthwork'. Surname extinct 15th c. **De Castel** (**de Castro**). hue q.v.
Castles	see **Chateaux**.
Castor, de	field: le Rocher, SM. Origin unknown.
Câtelier, le	land: N of l'Hyvreuse, SPP; E of l'Ancresse, CduV. Probably 'minor earthwork'. Recorded **Quatelier** 1591. Misspelt **Câtelain**, 20th c.
Câteline, la **Vallée**	land: L'Erée, SPB. Surname extinct 15th c. **Cateline**.
Catherine, **Ste**	house: la Ramée, SPP. 18th c. house name.
Catière	see **Quatière**, la.
Câtillon, le	house, land, road: N of Rocquaine, SPB. 'minor earthwork'.
Catioroc, le	land: W of Perelle, SSV. Former watch house. 'earthwork on the rock'. Also spelt **Castiaurocq**, 1534. garde q.v.
Catonnières, les	land: Jerbourg, SM. Origin uncertain.
Cauchebrée, de	house, land: les Maingys, VdelE. Surname extinct 15th c. **Caucheobrunne** (spelt **Coche brais**, 1700).
Cauchez, au	road, fields: rue Cauchez, SM. Surname extinct 17th c. **Le Cauchez**.
Cauchie, la	causeway, harbour: SPP. 'pier'. Medieval construction.
Caudré, le	see **Coudré**, le.
Cauf, les **Buttes** au	field: le Courtil Ronchin, SSV
Caufestrain, de	fields: les Hougues, C. Origin unknown.
Caugneron, le	see **Cogneron**, le.
Caunafle, de	land: la Lague, SPB. Surname extinct 15th c. **Caynaffle**.
Caunuette, la	fields: la Croix Foret, F. Origin unknown.

Cauvains, les	house, field: les Girards, C. Surname extinct 15th c. **Cauvain**, also spelt **Covin**. Also feudal holding of land, Fief **Covin**, C.
Caux, aux	fields: les Picques, SSV; les Paas, SPB.
Caux, les **Rocques** au	land: les Cambrées, SPB. Surname extinct 17th c. **Le Cauf**.
Cavaleur, le **Long**	see **Avaleur**, le long.
Cène, la **Croûte** au	fields: W of les Domaines, SSV/SPB. Possibly from rente owed for church communion service. croûte q.v.
Censière, la	fields: N of les Eturs, C; Plaisance, SPB; les Blicqs, les Naftiaux, SA. Land owing 'cens' (pence) for reclamation, also spelt **Sensière**.
Censière, la grande	land: near les Queux, C.
Cerceuil, le	see **Serqueur**.
Cevallet, le	fields: la Hougue Anthan, SPB; le Grand Menage, T. Origin unknown. Also spelt **Scevallet**.
Cévecq, au	fields: N of le Clos Vivier, SSV. Origin unknown.
Cévecq, rue du	road: unlocated, SSV.
Chailles, de	see **Cailles**, les.
Chaise au **Prêtre**, la	see **Prêtre**, la **Chaise au**.
Chambre, la	house, land: now called Le Bigard, F. Origin unknown. Earliest reference 1671.
Chambrette, la	land: Clos Vautiers, S of Grande Mare, C. 'pound'.
Champ(s), les	see **Camp(s)**, les.
Champhuet, le	see **Huet**, le **Champ**.
Chanteur, **Mare** au	field: la Rousaillerie, SPP. Surname extinct 16th c. **Le Chanteur**.

Chapelle (f)	medieval chantry chapels and chapels of ease listed by parish. See also individual entries.
SPP	St **Jacques** St **Jehan du Bosq** St **Julien** du (**manoir de**) **la Grange** Ivy Gates **de la Lorette** du **Franc Manoir au Marchant**
SSP	St **Clair** St **Thomas** (**Manoir d'Anneville**) **Notre Dame des Marais**
CduV	St **Magloire** (St **Mallière**)
VdelE	**Notre Dame de l'Epine**, **Notre Dame du Pulias**
C	St **George**, St **Germain**
SSV	Ste **Apolline**
SPB	St **Briocq**
T	none
F	none
SM	St **Jean**, **Notre Dame de la Houguette**
SA	none
Chapelle, la	fields: les Beaucettes, CduV. Presumably adjacent or connected to former chapel of St Magloire, q.v.
Chapelle de **St George**, Fief de la	feudal holding of land, St George, C.
Chapelle, la **Longue**	field: Paradis, CduV. Near former chapel of St Magloire.
Chapelle, la rue de la	road: Pont Perrin, VdelE. After **Chapelle Notre Dame de l'Epine**. Also called rue des Priaulx. q.v.
Chapinnière, la	land, harbour: SSP. Surname extinct 17th c. **Chapin**.
Charde(**t**), la **Rocque**	fields: Hougue Juas, CduV. Extinct surname **Chardronet**. See **Cardronet**. Earliest reference 1654.
Chardons, ès	field: les Caches, C. 'teazels'.

Chardronet, le	see **Cardronet**, le.
Chariot, la rue au	road, field: Bon Port, SM. 'cart'. Also spelt **Carriot**.
Charles, route	road: modern 20th c. road, SPP.
Charriere, la **Blanche**	field: la Rochelle, CduV.
Charrière, la **Neuve**	see **Querrière**, la **Neuve**.
Charrières, les **Blanches**	fields: les Paysans, near la Pomare, SPB. Permanent stony cart track (at side of field), presumably made with 'white' quartz and mica. Recorded **Blanche Querrière**, 1591. Often confused with carrière q.v. cache, chasse, querrière, traversain q.v.
Charroterie, la	district: la Charroterie, SPP. 18th c. 'cart sheds'. Also spelt **Charreteterie**, **Carterie** and **Querterie**.
Charruée, la	houses, land: W of les Landes du Marche, VdelE/C. 'ploughed'. Derived from superficial land measure, caruée (corvée). Also spelt **Carruée**, **Querruée**, la **Querrière**. corvée q.v.
Charterie, la	land, fields: all parishes. 'cart shed'. Also spelt **Carterie**, **Querterie**. See **Charroterie**.
Chasse Vassal, rue	road: see **Ann's** place, **Vassal**, la **Chasse**. fontaine q.v.
Chataigner, le	land: Grange, SPP. Origin uncertain.
Château (m)	'castles'. château **Cornet**, SPP chateau des **Marais**, SSP château du **Valle**, CduV château de **Rocquaine**, (replaced by Fort Grey) SBP château **de la Corbiere**, F château **de Jerbourg**, SM
Chaumette, la	house, land: les Landes, F. 'thatched'. also **Chaumière** (19th c. usage)

Chaussée, la	see **Cauchie**, la.
Chavatte, le **Clos**	land: rue Sauvage, SSP. Origin uncertain.
Chechire, la **Hougue**	field: les Beaucettes, CduV. Surname extinct locally 16th c. **Chechire**. (15th c. English immigrant family). hougue q.v.
Chef Bordage, le	feudal holding of land, N of le Grantez Mill, C. Bordage q.v.
Chèliziers, les	land: Pleinheaume, VdelE; la Pomare, SPB. Possibly 'wild cherries'.
Chemin (m)	road, all parishes. 'highway', i.e. any public way or thoroughfare. Route (route), Rue (road, street), ruette (land) q.v.
Chemin de l'**Eglise**, le	roadway, all parishes. Way to the parish church.
Chemin le **Roi**, le	roadway, see **Roi**, le **Chemin le**.
Chemin, camp du **Grand**	see Chemin, le Grand.
Chemin, le **Bas**	fields: les Bas Courtils, SSV. 'lower highway'. Also spelt **Bas Quemin**.
Chemin, le **Grand**	fields: les Gravées, les Godeines, SPP; route de Jerbourg, SM. 'main highway'.
Chemin, le **Haut**	house, fields: la Couture, SPB. 'upper highway'. croix q.v.
Chemin, le **Neuf**	house, fields: le Neuf Chemin, SSV. Original road now in reservoir. 'new highway'.
Chemin, le **Profond**	road: le Haut Chemin, SPB. 'steep highway'.
Cheminée, la **Ronde**	houses, land: S of Nocq road, SSP; W of les Goubeys, VdelE. 'round chimney'.
Chemins, les **Quatre**	houses: W of Baugy, CduV. 'four highways'.
Chêne, le	district, fields: le Chêne, F. 'oak'.
Chêne, le **Camp** au	land: le Grand Belle, SA. 'oak' on strip of land.

Cherf(**s**), le(s)	houses, fields: S of Cobo, C. Surname extinct 17th c. **Le Cherf**. Also feudal holding of land, Fief des **Cherfs**, C.
Cherfs, hougue des	land: near les Pins, C. Derivation as above.
Cherfs, la **Saudrée** des	land: les Cherfs, C. 'willow patch'. Derivation as above. Saudrée q.v.
Cherfs, ruette des	land: les Corneilles, C. Derivation as above.
Cherverie, la	house, land: la Gelé, C. From surname **Le Cherf**, see above.
Cheval, la **Rocque** au	land: les Terres, SPP. 'horse rock'. Origin unknown (near rocque ès **Chèvres**).
Chevalier, le	field: le Mont Frie, SM. Surname **Le Chevallier** (extant in Jersey). Also feudal holding of land, Fief au **Chevallier**, C.
Chevallots, les	land: E of les Gravées, SPP. Presumably extinct surname.
Chèvres, la **Rocque** ès	land: les Terres, SPP. 'goat rock'. (near rocque au **Cheval**).
Chien, la **Fosse** au	land: le Gron, SSV. Origin unknown. Recorded 1663. Possibly dog-leg shaped.
Chien, la **Pièce** à	land: N of Saumarez park, C. Origin unknown. Recorded 1663. Possible derivation as above.
Chien, la rue à	road: les Annevilles, SSP. Also called Rue de la Cache. Recorded late 19th c. Possible derivation as above.
Chinchon, **Mont**	fields, headland: N of le Catioroc, SSV. Surname extinct 16th c. Recorded **Chinchon** 1536. Also 18th c. battery. mont q.v.
Chivret, de	fields: les Arguilliers, VdelE; les Landes, C; l'Alleur, T; les Bruliaux, F. Surname extinct early 19th c. **Chivret**, also spelt **Chyret**.
Chivret, de **David**	field: les Rohais, SPP. Derivation as above.
Chivret, de **Guillaume**	field: S of le Crôlier, T. Derivation as above.

Chivrette, la	field: les Rohais, SA. Derivation as above.
Choffins, douit des	stream: les Choffins, SSV. Surname extinct 15th c. Choffin.
Choffins, les	houses, fields, road: les Choffins, C/SSV. Derivation as above. vallée q.v.
Choisi, de	house: les Gravées, SPP. 'chosen'. Early 19th c. house name.
Chouet, de	headland: W of Mont Cuet (tip), CduV. Surname extinct 15th c. **Chouet**, **Cuhouel** (**Pierre Chouet**). Possible the name has a connection with the Vale Priory rolls, c. 1330. Crevel, le Mont, Cuet, le Mont q.v.
Churches, **Parish**	see **Eglises Paroissiales**.
Cimetière des **Frères**, le	cemetery: E of Elizabeth College, SPP. Formerly garden and cemetery of a small house of Franciscan Friars before the Reformation. Frères q.v.
Cimetière, le	land: le Hamel, VdelE; St Briocq, SPB. Burial ground not attached to a church. chapelles q.v.
Cinq Verges, les	see **Verges**, les **Cinq**.
Cité, la	house: le Grais, SPB. First reference 1756, origin unknown.
Clair, **Saint**	house, land: St Clair, SSP. See St Clair. Medieval chapel (site cleared). chapelle q.v.
Claire Mare, la	see **Mare**, la **Claire**.
Clais, la	fields: les Houmets, C; les Bas Courtils, SSV. Origin uncertain, possibly 'hurdle'. Also spelt **Claye**. clef q.v.
Clanque, les **Camps**	land: Sous l'Eglise, SSV. Extinct surname **Jean Esclenque**. Earliest reference 1663. camps q.v.
Claquez, à	land, field: Candie, C. Probably extinct surname.
Clarion, le **Clos**	land: les Marais, SSP. Origin unknown.

Clef, de **Cour**	field: S of le Douvre, T. Origin uncertain, possibly 'small hurdle'. clais q.v.
Clerc(q)s, les	houses, fields: ruettes Brayes, rue a l'Or, SPP; N of Bordeaux, CduV; rue des Marais, VdelE; le Villocq, C; Contrée des Clercs, la Carriere, SPB. Surname extinct 17th c. **Le Clerc** but now present again, 20th c.
Clercqs, le **Pont** ès	land: les Mauxmarquis, SA. Derivation as above.
Clerel, le	field: la Fallaize, Jerbourg, SM. Origin unknown. Also spelt **Clairet**.
Clerelle, la	field: l'Alleur, T. Origin unknown. Recorded **Clareve**, 1595, **Clerrelle** 1644.
Clifton, de	district: SPP. Early 19th c. name chosen by Pierre Mainguy. Cliff town, town on a cliff.
Cloison, le	field: St Clair, SSP. 'partition'.
Clopains, les	see **Pains**, les.
Clores, les	field: Grange, SPP. Origin uncertain.
Clos, le	houses, fields, all parishes. 'enclosure'. closel, clôture, enclos q.v.
Clos	with descriptive names are listed below.
Clos au Pommiers, la rue des	road: unlocated, T.
Clos d'**Hys**, le	see **Théhy**, le Clos.
Clos Landais, rue des	road: N of Les Islets, SPB.
Clos, le **Grand**	house; la Charruée, C. 'large enclosure'. Recorded 1270.
Clos, le **Neuf**	houses, fields; la Haize, CduV; les Genats, C; la Vieille Rue, le Neuf Clos, SSV; Rocque Poisson, SPB. 'new enclosure'.
Clos, rue le	road: S of les Martins, SSP.

Clos with attached proper names. See also individual entries.

SPP	Clos **Fabien**, Clos **Vaudins**.

SSP	Clos **des Sept Rocques**.
CduV	Clos **au Jonc**, Clos **Moulant**, Clos **Pains**, Clos **Allix de Lande**.
VdelE	Clos **Drouin**.
C	Clos **Dumaresq**, Clos des **Fourches**, Clos **Frazel**, Clos **Gervaise**, Clos **du Groignet**, Clos **au Gros**, Clos **à l'Herbe**, Clos **d'Hys** (Théhy), Clos **au Jonc**, Clos **Lengier**, Clos **Marinel**, Clos **ès Meches**, Clos **de Paré**, Clos **au Pelley**, Clos **Piton**, Clos **au Préel**, Clos **Quartier**, Clos **au Romil**, Clos **à Ronces**, Clos **de la Salle**, Clos **Simon**, Clos **de la Vallée**, Clos **Vautier**, Clos **au Veaux**, Clos **Verron**, Clos **Vichet**, Clos **au Vietiu**, Clos **Vivier**.
SSV	Clos **à la Dame** (du Fief), Clos **Drouins**, Clos **d'Enfer**, Clos **à l'Herbe**, Clos **Hoguet**, Clos **Huchon**, Clos **Jamin**, Clos **à Jenets**, Clos **Landais**, Clos **Mansell**, Clos **Massy** (Le Dien), Clos **du Mont** (au Nouvel), Clos **de Porte**, Clos **Samson**, Clos **Vivier**.
SPB	Clos **Mabire**, Clos **Poisson**, Clos **du Puits**, **Neuf** Clos **de la Rocque Poisson** (Paisant).
T	Clos aux **Pommiers**.
F	Clos **Poidevin**, Clos **Richard**, Clos **Bernard**.
SM	Clos **de l'Etac**, Clos **Maze**, Clos **Piquet**, Clos **Querette** (Carette), Clos **au Barbier**.
SA	Clos **au Foin**, Clos **Hunguer**, Clos **Lucas**.

Closel, le — houses, fields: le Grève, CduV. 'small enclosure'. Diminutive of clos, q.v.

Closemare, de — field: la Bâte, CduV. 'pond enclosure'. Recorded **Clode Mare** 1591.

Closios, les — fields: le Jamblin, CduV. Plural of closel q.v.

Clospains, les — see **Pains**, le **Clos**.

Clôture, la — houses, fields, all parishes. 'large enclosed area'. Clos, closel, closios q.v.

Cloture, la grande — land: la Corbière, F.

Clouet, de — land, fields: le Bouet, SPP; les Fauquets, C; les Norgiots, la Contrée de l'Eglise, SA .

	Surname extinct 17th c. **Clouet**. Misspelt **Clasnest**.
Clungeonnerie, la	site of Trinity Church, SPP. Purchased 10.12.1794. Extinct surname **Clungeon**. Recorded **Pierre Clungeon**, 1581. Trinité q.v.
Cobehans, les	land: unlocated, T. Origin unknown. Recorded 1511.
Côbo, de	houses, fields: la Ruette, SSP; Côbo, C. Surname extinct 16th c. **De Caubo**. Recorded John De Caubo, 1479.
Côbo, la **Mare**	land: S of Mont Cuet, CduV. Derivation as above.
Côbo, le **moulin** de	mill: unlocated, C. Derivation as above.
Côbo, le **Puits** de	spring (fontaine): la Lande, C. Add Derivation as above. mare, puits q.v.
Côbo, route de	road: formerly rue Ysorey. Orais q.v.
Cobois, Fief des	feudal holding of land. Dependency of Fief des Vingt Bouvées du Villian Fief le Comte, C.
Cocagne, de	district: N of Bordeaux, CduV. Origin unknown.
Cochon, le	field: la Croûte, SSV. Presumably extinct surname **Coschon**.
Cocq chante, la **Rocque** où le	field: le Dos d'Ane, C. Earliest reference 1614.
Cocq, **Coq** au	field: les Paas, SPB.
Cocq, le **Val** au	fields: rue Piette, C. Also feudal holding of land, Fief au **Cocq**, C.
Cocq, les **Camps** au	fields: le Hauteur, SPB.
Cocquerel (**s**), le(s)	houses, fields: la Grande Maison, VdelE; les Hougues, les Cocquerels, C; les Lohiers, SSV.
Cocquerel, de **Pierre**	field: S of le Hauteur, SPB. Surname extinct 17th c. **Cocquerel**.
Cocquerel, fosse	land: unloc. SPB.

Cocquerie, la	house, field: la Terre Norgeot, SSV. Extant surname **Le Cocq**.
Code, le **Camp**	fields: les Bas Courtils, SSP. Surname extinct 16th c. **Gode**. Misspelt **Code** 19th & 20th centuries. camp q.v.
Coemil, le	fields: E of les Laurens, SPB/T. Origin unknown. Also spelt **Coemy**, **Cocmul**, (Extente 1331).
Coffinot, le	land: Elizabeth college, SPP; le Carrefour, C. 'sliver' of flat land. Also spelt **Coffinotte**.
Cogneron, le	land, fields: les Etibots, SPP; Cognon, CduV; le Sauchet, T; les Villets, F; la Quevillette, la Fallaize, SM; rue Marquand, SA. 'recessed corner of field'. Also spelt **Cognon**, **Caugneron**.
Cognon, le	former field: E of Electricity Power Station, CduV. Probably from **Cogneron**, above.
Cohu, d'**André**	field: les Rohais, SA.
Cohu, de	fields: le Haut Pavé, C. Extant surname **Cohu**.
Cohu, la rue	road: Saumarez park, C. Recorded 1470. Derivation as above. Also feudal holding of land, Fief **Cohu**, C.
Cohue, la	site: la Plaiderie, SPP. Former Royal Court House. Plaiderie q.v.
Coignet, le	fields: la Croûte Becrel, CduV; Pleinheaume, VdelE; les Landelles, C; la Porte, SSV. Diminutive of coin, q.v.
Coin, (m)	'corner'. Descriptive names are entered under the substantive name; coin **Casson**, coin **Fallaize**, coin **Colin**.
Coin Cassot, le	see **Cassot**, le **Coin**.
Coin Colin, le	see **Colin**, le **Coin**.
Coin Fallaize, le	see **Fallaize**, le **Coin**.
Coismil, **Coismy**	see **Coemil**.

Colborne road — road: les Vardes, SPP. Named after Major General Sir John Colborne, Lieutenant Governor, 1825-28.

Coliche, de — field: les Poidevins, SA. After **Colliche Prey** (recorded 1701).

Colin, de — land: les Damouettes, SPP; les Pelleys, C; les Blicqs, SA. Forename, **Colin**. Also spelt **Collin**.

Colin, houguette — land: W of le Grand Fort, SSP.

Colin, la **Rocque** — land: le Gain, SPB. Derivation as above.

Colin, la rue — road, land: les Trachieries, SSP.

Colin, la **Croix Jehanne** — site: les Capelles, SSP. Former wayside cross.

Colin, le **Coin** — land: les Blanches, SM. After **Collin Blanche**.

Colin (**Collas**) **Martin**, la croute — land: les Blanches, SM.

Collas, de — fields: la Ronde Cheminée, SSP; les Jetteries, SSV. Forename or surname, **Collas**, from Nicolas.

Collas Brehaut, de — see **Brehaut**, de **Collas**.

Collas Des Mares, de — see **Mares**, de **Collas Des**.

Collas Roton, de — see **Roton**, de **Collas**.

Collenette, de — fields: Candie, C; la Grande rue, SSV. Forename (f) or later surname **Collenette**.

Collet, de — fields: le Haut Pavé, C. Surname extinct 15th c. **Collet** (extant Jersey).

Colleton, de — see **Colton**, de.

Collette, la — fields: les Buttes, SPB; la Croix Forêt, F. Forename (f) **Collette**.

Collette Nicolle, les **Camps** — see **Nicolle**, les **Camps Colette**.

Collin Denis, hougue	land: N of Vazon near Ft. Hommet, C.
Colomb, au	field: les Ruettes, SA. Surname extinct 16th c. **Coulomp**.
Colomb, le **Camp** au	land: le Felcomte, SPB. Derivation as above. camp q.v.
Colomb, le **Val** au	field: W of les Choffins, SSV; le Felcomte, SPB. Derivation as above. Misspelt **Colombre**, 19th & 20th centuries. val q.v.
Colombelle, la	house, fields: ruettes Braye, SM. Origin unknown.
Colombier, le	house, dovecot, fields: le Colombier, T. 'dovecot'. Originally 13th c. Rebuilt 1954. Belonged to Fief Janin Bernard (part of Fief au Canely).
Colombier, le	dovecot (ruined): Lihou. Belonged to priory of Lihou.
Colombiers, les	land, field: l'Islet, SSP; W of les Laurens, T. (no evidence of dovecots).
Colons, la **Fontaine** ès	field: le Rocher, SM. Surname extinct 15th c. **Colon** or **Coullon**.
Colton(s), des	fields: la Devise, SPB; la Fallaize, SM. Surname extinct 17th c. **Colleton**. Also feudal holding of land, Fief des **Coltons**, T.
Commercial Arcade	precinct of shops: off High street, SPP. Developed by James & George Le Boutillier, 19th c.
Communes, les	'commons'. Historically –

Communes de l'Ancresse	l'Ancresse, CduV
Communes des Landes	les Landes, CduV
Communes de la Rocque au Cucu	le Lorier, CduV
Communes de la Marette	les Romains, SSP
Communes de Pleinmont	Pleinmont, T
Communes de la Forêt	cliff land, F

Communes de Saint Martin — between Petit Bot and Petit Port, SM

Comte(s), les — houses, land, fields: les Comtes, SSV; la Claire Mare, SPB; Le Camp Tréhard, SA.

Comte, du **Fief** le — see **Felcomte**, le.

Comte, hougue au — land: rue de Cobo, C.

Comte, le **Clos** au — house, fields: le Clos au Comte, C. Recorded 1470.

Comte, les **Maisons** au — houses: W of Baugy, CduV. Extant surname **Le Comte**, **Le Conte**. (**Pierre Le Conte**, 1504). clos, maisons q.v. Also feudal holding of land, Fief **le Comte**, C/SSV/SPB,after the Earls of Chester who held the fief in the middle ages.

Conichon, le — land: les Nicolles, SSP. 'corner'.

Connerettes, les — field: S of le Pezeril, CduV. Origin unknown.

Connillier, la **Fosse** au — land: les Nicolles, F. Presumably extinct surname. Recorded **Cannelier**, 1671. fosse q.v.

Contrée Croix, la — houses: E of Mansell street, SPP. Cross street between Mansell street and le Bordage. croix q.v.

Contrée, la — district of houses, all parishes. territoire q.v.

Contrée de l'**Eglise** — see **Eglise**, l'.

Contrée des **Clercs** — see **Clercs**, les.

Contrée Mansell, la — see **Mansell**, de.

Contrôle, le — house: rue Berthelot, SPP. Origin unknown.

Coq, le — see **Cocq**, le.

Coquerel(s), le(s) — see **Cocquerel**.

Coqueterres, les — fields: la Maison de haut, SSV. After **Guillaume Cauqueterre** recorded 1270.

Corbe, la — field: les Bas Courtils, SSP. Origin unknown.

Corbel, de	land, fields: King's road, N of le Bouet, SPP; les Marais, SSP; les Jaonnets, SSV. Extant surname **Corbel**, also known since early 19th c. as **Corbet**.
Corbel, la **Vallée**	land: les Niaux, C. Derivation as above. vallée q.v.
Corbière, **Fort**	see **Corbière**, le **Chateau**.
Corbière, la	promontory, district: S of Havelet, SPP; la Corbière, F.
Corbière, le **Château** de la	promontory: fortification, la Corbière, F. Origin unknown. château q.v.
Corbillons, les	fields: E of les Padins, SSV. Surname extinct 17th c. **Corbillon**.
Corbin(s), les	land: les Terres, SPP, la Hougue Fouque, SSV; les Islets, SPB. Extant surname **Corbin**.
Corbin, la **Croûte**	land: le Vauquiedor, SM. Derivation as above. croûte q.v.
Corbin, la **Fontaine** ès	land: les Vaulorens, SPP. Derivation as above. fontaine q.v.
Corbin, le **Camp**	land: les Picques, SSV; les Adams, SPB. Derivation as above. camp q.v.
Corbin, l'**Ecluse**	land: les Poidevins, SA. Former mill pond. Derivation as above. ecluse q.v.
Corbin, le **Ménage**	land: les Adams, SPB. Derivation as above. ménage q.v.
Corbinerie, la	house: Oberlands, SPP. 20th c. house name.
Corbinez, les	houses, district: la Palloterie, SPB. Surname extinct 14th c. **Corbinetz**. Also feudal holding of land, Fief de **la Corvée ès Corbinets**, SPB.
Cordage, le	houses, land: les Amballes, SPP; la Moye, CduV; le Carrefour, C; les Domaines, le Hurel, SSV; rue St Pierre, SPB; Calais, les Maindonnaux, SM. 'rope walk'.
Corderie, la	see **Cordage**.

Cordiers, les — houses: la Couperderie, SPP. Surname extinct 15th c. **Cordier**.

Corneille(s), les — house, fields: les Corneilles, les Baissières, C; St Briocq, SPB. Surname extinct 16th c. **Corneille**. Also feudal holding of land, Bouvée des **Corneilles**.

Corneille, la hure — land: la Trappe Berger, C.

Cornet, le **Bordage** — street: le Bordage, SPP. Surname extinct 15th c. **Cornet**. Bordage q.v.

Cornet, le **Château** — fortification: Castle Cornet, SPP. Derivation as above. Château q.v.

Cornets, la rue des — road: Cornet street, SPP. Derivation as above. rue q.v.

Corniau, de — field: rue à l'Or, SPP. Origin unknown.

Cornicheries, les — houses, land: les Cordiers, SPP. Origin uncertain.

Cornières, à trois — fields: le Bouet, les Croûtes, SPP; unlocated SSP; les Landes, le Closel, CduV; near la Haye du Puits, C; la Hougue Fouque, SSV; les Paysans, SPB; le Planel, T; les Mauxmarquis, SA. triangular shaped fields.

Cornillion, de — land: unlocated, SPB. Surname extinct 15th c. **Cornillot**.

Cornis, de — field: le Vauquiédor, SPP. Origin unknown.

Cornu(s), les — district, fields: le Bouillon, SSV; les Cornus, SM. Extant surname **Le Cornu**.

Cornu, de **Helier Le** — field: les Bémonts, SA

Corvallet, le — field: les Sommeilleuses, F. 'short small valley'.

Corvée, la —
1. land measure. 12 bouvées (240 vergées).
2. land from which materials were taken for the corvée (compulsory road repairing).

Corvée(s), les — fields: les Marais, SSP; la Maison de bas, CduV; les Barrats, VdelE; les Grands

	Moulins, la Masse, C; ruette Suart, SPB; Pleinmont, T; S-E of les Blanches, SM; les Mourants, les Rohais, SA. Land from which materials were taken for la corvée, compulsory road repairing required of landowners. The obligation replaced by tax in 19th c.
Corvées, camp des	see **Corvee(s)**, les.
Corvées, fief des	feudal holding of land. Dependency of Fief des Vingt Bouvées du Villain Fief le Comte, C.
Corvées, hougue des	land: E of Sohier road, CduV.
Corvées, ruette des	lane: E of la Hougue Ricart, CduV.
Costant, le	land: N of les Monts, SSP; Bon Port, SM. Origin unknown.
Costin, de	land: les Rouvets, VdelE. Origin unknown.
Cote ès Ouets	house: les Rouvets, VdelE. 19th c. nickname, 'cackling geese'. Moutonnerie q.v.
Cotentin, le	fields: le Hurel, SM. Presumably extinct surname. Recorded ès **Cotentins**, 1488.
Cotentins, fontaine	spring: les Courtes Fallaizes, SM.
Côtes aux **Monts**, les	see **Aumont**, la Cotte.
Côtil, de **Thomas**	land; les Terres, SPP. Surname extinct 16th c. **Costil**, terre q.v.
Côtil Mustel, le	land: les Talbots, C. 'steep slopes'.
Côtillon, le	land, all parishes. Diminutive of côtil. 'small steep slope'.
Côtils de **Glategny**, les	land: E of l'Hyvreuse, SPP.
Côtils, les	land, fields, all parishes.
Cou-cou, le	see **Cucu**, le.
Coudré, le	district, fields: E and uphill of Rocquaine, SPB. Possibly extinct surname **Caudre**. Hoirs **Jehan Coudroy**, 1584.

Coudré, la rue du	road: formerly via les Fontenelles, le Câtillon, SPB. Derivation as above.
Coudré, le **Moulin** du	site: le Coudré, SPB. windmill, pre-1671. Demolished 1941. Derivation as above. Moulin à vent q.v.
Couepel, le	field: les Clospains, CduV. Origin unknown. Earliest reference **Coispel**, 1727.
Counuette, la	field: unlocated. F. Origin unknown.
Coupée Doublier, la	see **Doublier**, la **Coupée**.
Couperderie, la	house: les Cordiers, SPP. 18th c. house name, after Mr Cooper who lived there for a short time.
Cour Clef, le	see **Clef**, le **Cour**.
Cour, de la	field: le Houmtel, CduV. After **Thomas De La Cour**, living in 1591. Surname extinct 18th c. **De La Cour**.
Cour de Longue, la	houses, fields: la Cour de Longue, SSV. Adjacent to court seat of Fief de Longues. Court seat at friquet adjacent to above.
Cour, de **Thomas de la**,	fields: le Chene, F. 19th c. court seat at le Frie Plaidy, C. Also feudal holding of land, Fief **de la Cour**, C.
Courchières, les	fields: les Grippios, CduV; les Adams, SPB; le Coin Fallaize, SM. Origin uncertain. Also spelt **Coursières**.
Courly, la **Fontaine** au	field: la Grande Lande, SSV. Recorded 1695. Surname extinct 15th c. **Courlu**.
Courly, la **Fosse** au	field: les Beaucamps, C. recorded 1551. Derivation as above.
Cour Ricart, de la	see **Ricart**, la **Cour**.
Cour Ricart, fief de la	feudal holding of land. SPB.
Cours Camps, les	see **Camps**, les **Cours**.
Coursières, les	see **Courchières**.
Court(e)s Fallaize, la	see **Fallaize**, les **Courtes**.

Court, fontaine	spring: Bon Port, SM.
Courtes Pièces, les	see **Pièces**, les **Courtes**.
Courtil, le	field, all parishes.
Courtil(s)	with descriptive names are listed below:
Courtil à Paix, le	field: les Annevilles, SSP. Paix q.v.
Courtil au Préel, le	field: le Bouillon, SA. Preel, q.v.
Courtil Blicq, le	field: les Rouvets, SSV. Blicq q.v.
Courtil Blicq, le	houses, field: les Blicqs, SA. Blicq q.v.
Courtil Blondel, le	field: la Pomare, SPB. Blondel q.v.
Courtil Brehaut, le	field: le Hurel, SM. Brehaut q.v.
Courtil Carré, le	'square' field, SPP; C; SSV; SPB.
Courtil d'aval, le	'lower' field, all parishes.
Courtil de **Bas**, le	'bottom' field, all parishes.
Courtil de **Devant**, le	'front' field, SSP; C; SSV; SPB; SM.
Courtil de **Haut**, le	'top' field, SSP; CduV; SM; SA.
Courtil de **l'Eglise**, le	field: les Rohais, SPB. Eglise q.v.
Courtil de **Lisle**, le	field: E of les Brehauts, SPB. Lisle, De q.v.
Courtil de **Paul**, le	field: le Lorier, SPB. Paul q.v.
Courtil Guilbert, le	field: la Palloterie, SPB. Guilbert q.v.
Courtil le **Vieux**	land: Roland Road, SSP; S of Paradis, CduV; La Pomare, SPB; le Douit de la Porte, SM.
Courtil Pierre Samson, le	field: rue de Bouverie, C. Samson q.v.
Courtil Pointu, le	'pointed' field, SM.
Courtil Prochain, le	'next' field, SPB.
Courtil Ronchin, le	field: les Gervaises, SSV. Ronchin q.v.
Courtil(s), les **Grand(s)**	fields: les Trachieries, SSP; le Neuf Clos, SSV; la Lague, T; N of le Vauquiedor, SA. 'big' field.

Courtil, le **Neuf**	fields, all parishes. 'new' field, i.e. a more recent enclosure.
Courtil, le **Plat**	fields: le Hurel, S of les Fauconnaires, SA. 'flat' field.
Courtil, le **Profond**	field: la Hougue du Moulin, CduV. 'steep ' field.
Courtil, le **Vieil**	fields: les Fontenelles, SSV; la Pomare, SPB; les Landes Hue, T. 'old' field, i.e long enclosed.
Courtillet(s), le(s)	houses, fields, all parishes. Diminutive of courtil. 'small' fields.
Courtils, les **Bas**	fields: les Banques, SSP; E of le Gron, SSV. 'low' fields, relative to the area.
Courtils, les **Hauts**	fields: N of la Maison de bas, VdelE. 'top' fields.
Courtils, **Sous** les	land, fields: Albecq, C
Cousin, croûte au	land: unlocated. C. Surname extinct 17th c. Cousin.
Cousin, de	fields: le Passeur, CduV; les Fries, C. Derivation as above.
Coutanc(h)e, de	house, fields: la Maraitaine, CduV; les Grands Moulins, C. land: Baugy, CduV. Surname **Coutanche** (**Coutanches**).
Coutance, rue	house, fields: la Maraitaine, CduV; les Grands Moulins, C. Derivation as above.
Coutanchez, les	house, fields: N of la Rousaillerie, SPP. Reference **Louys Le Coustanches**. Recorded 1574. Derivation as above.
Couteur, au	field: Sohier, CduV. Recorded **rue au Cousteur**, 1591. Extant surname **Le Couteur**.
Couture(s). les	houses, fields: la Couture, SPP; E of le Tertre, N of Vale school, CduV; le Friquet, C; la Couture, SPB; les Sages, T; les Coutures, les Caches, SM. 'cultivated'.

	Also feudal holding of land, Fief **Couture**, SPB.
Couture, rue de la	road: les Sages, T.
Covin, Fief	feudal holding of land. Dependency of Fief de Carteret, C.
Covins, les	see **Cauvains**, les.
Covins, rue des	road: N of the Castel Hospital. Cauvains, q.v.
Crabes, les	fields: Perelle, SSV. Surname extinct 15th c. **Crabe**. Misspelt **Crabbes**, 20th c.
Cras, au	fields: les Grandes Rues, SPB; les Blicqs, SA. Extant surname **Le Cras**.
Cravellet, le	fields: la Bouvée, SM; les Vauxbelets, SA. Origin unknown.
Crespel, la **Rocque**	field: le Coudré, SPB. Surname extinct 14th c. **Crespel**. rocque q.v.
Crête, la	fields: les Monts Hérault, SPB; le Douvre,T. 'crest'.
Creux ès Fées, le	see **Fées**, le **Creux** ès.
Creux Mahie, le	see **Mahie**, le **Creux**.
Creux, le	land: les Monts Hérault, SPB. 'hollow' on cliff.
Crève Cœur, de	land, road: le Pezeril, CduV; N of Castel church, C. 'heart break'. i.e. unyielding soil.
Crève Cœur, la **Croix**	land: W of le Frie Baton, SSV. Derivation as above. croix q.v.
Crevel(t), le **Mont**	promontory, fields: E of Bulwer Avenue, Southside, SSP. ('T' added by Ordnance Survey surveyors in 1900.) Surname extinct 15th c. **Crevel**.
Crevel, le **Mont**	promontory: le Mont Cuet (tip), CduV. (Named Mont Cuet by Ordnance Survey surveyors in 1900). Derivation as above. See Chouet.

Criquettes, les	field: N of route du Braye, CduV. Origin unknown. Recorded **Criquettes**, 1591. Misspelt **Griquettes** 1922.
Crochon, le	field: le Vacheul, F. Surname extinct 15th c. **Le Crochon**.
Crochonnière, le	fields: le Courtil Guilbert, SPB. Derivation as above. Also feudal holding of land, Fief au **Crochon**, SPB.
Crocq, le	houses, land: South side, SSP; la Marette, SSV; Rocquaine, SPB. 'low headland';
Croisée, la	house, fields: N of les Tielles, T; S of le Bourg, F. Origin uncertain. Also spelt **Croisi**.
Croissillon, le	land: rue de l'Arquet, SPB. 'small cross'. croix q.v.
Croix, la	land, all parishes. Former wayside crosses, memorials to the dead, and places for votive offerings and prayer.
Croix de **Pierre**, la	cross: les Vallées, C. 'stone' cross.
Croix du **Bois**, la	cross: E of la Maraitaine, CduV. 'wooden' cross.
Croix Guillemette, hougue de la	land: le Dos d'Ane, C.
Croix Rompue, la	cross: le Neuf Chemin, SSV. 'broken' cross.
Croix, la	approximate locations for croix with 'unattached' names. The precise location has been lost in most instances.
SPP	Glategny, Contrée Croix (Mansell), Havilland, La Ramée.
SSP	les Capelles, les Banques, la Ronde Cheminée, les Effarts.
CduV	none.
VdelE	none.
C	la Ville Baudu, Rocque au Mer, la rue Piette, les Landes, le Villocq.
SSV	les Jénémies, la Crève Cœur, le Gron.
SPB	L'Erée, les Adams, le Haut Chemin, les Marchez.

T	none.
F	none.
SM	la Bouvée, Icart.
SA	Les Naftiaux, les Fauconnaires.

Croix with proper names listed by parish, see also main text.

SPP	none.
SSP	Croix **Jehanne Collin**, Croix **Nez** (**Naye**), Croix **Plichon**.
CduV	Croix **Bernard** (Besnard), Croix **Damun Gouanne**.
C	Croix au **Beir**, Croix **St Germain**, Croix **Godefroy**, Croix **Nicolle**, Croix **Guillemette**, Croix de **Belle Voye**, Croix **Jean Toulle**, Croix au **Prestre**.
SSV	Croix **Martin**, Croix **Paysant**.
SPB	Croix **Brehaut**, Croix **Ivelin**, Croix **Perrinne**, Croix **Robin**, Croix **St Pierre**.
T	Croix **Dom(inus) Nicolas** (**Revel**), Croix **Jehans**, Croix au **Sage**.
F	Croix **Forêt**, Croix **Croquet**.
SM	Croix **Bertrand**, Croix **Fallaize**, Croix **Guerin**, Croix **Guillon**.
SA	Croix au **Baillif**.

Croix, le camp	land: Icart, SM.
Crôlier, la hure du	land: S of le Crôlier, T. Also known as Letat or Letac.
Crôlier, le	houses, fields: Pleinmont, T. Origin unknown.
Croquet, le croix	cross: la Soucique, F.
Croquet, le	land: le Bas Marais, le Mont Chinchon, SSV; l'Eree, SPB; la Soucique, F. Diminutive of **Crocq**. 'spur' of land.
Croquet, le	see **Crocquet**, le.
Croquets, camp des	land: near la Passée, SSP.
Crottey, le	field: les Fontenelles, SSV. Origin unknown, spelt **Croté**, 19th & 20th centuries.
Croûte, la	houses, land, fields, all parishes. 'crusty'

	heavy soil. Greatest area of croûte is in upper parishes.
Croûte de **Bas**, la	fields: Plaisance, C; la Hougue Fouque, SSV; les Maumarquis, SA. 'lower'.
Croûte de **Haut**, la	fields: les Grands Moulins, C; les Gouies, SA. 'upper'.
Croûte, la **Basse**	field: S of la Maison d'Aval, VdelE. 'low'. Rossiaux q.v.
Croûte, la **Grande**	field: W of le Gelé, C. 'large'.
Croûte, la **Haute**	fields: W of la Grande Maison, VdelE; la Hougue Fouque, SSV. 'top'.
Croûte, la **Ronde**	land: les Hougues, C; le Hurel, SM. 'round'.

Croûtes with proper names listed by parish, see also individual entries.

SPP	Croûte **Escolasse**, Croûte **à Genets**, Croûte **Gosselin**, Croûte **Guignon**, Croûte **Havilland**, Croûte **Herivel**, Croûte **de la Petite Marche**, Croûte **des Nicolles**, Croûte **Petiot**, Croûte **Piart**, Croûte **de Putron**, Croûte **des Rohais**.
SSP	Croûte **des Bigotez**, Croûte **Bouilleuse**, Croûte **Fontaine**, Croûte **France**, Croûte **de Marion**, Croûte **des Martins**, Croûte **du Mont Crevel**, Croûte **des Nicolles**, Croûte **des Perrins**.
CduV	Croûte **d'Anne Jouanne**, Croûte **Becrel**, Croûte **Beton**, Croûte **du Grand Camp**, Croûte **du Camp Rougier**, Croûte **de l'Ecole**, Croûte **dite Longue Raée**, Croûte **du Mauvaret**, Croûte **de la Miellette**, Croûte **Paris**, Croûte **du Pezeril**, Croûte **à la Pie**, Croûte **Pigny**, Croûte **du Puits**, Croûte **Quesnel**, Croûte **Rauf Salemon**, Croûte **Renouvet Hamelin**, Croûte **à Ronge**, Croûte **au Sage**, Croûte **au Vrée**.
VdelE	Croûte **des Bruniaux**, Croûte **Maingy**, Croûte **Mauger**, Croûte **au Mière**, Croûte **à Saulx**.
C	Croûte **au Beir**, Croûte **au Cousin**, Croûte

	des Emrais, Croûte **Gosselin**, Croûte **des Ravenios**.
SSV	Croûte **Blondel**, Croûte **au Cène**, Croûte **ès Dans**.
SPB	Croûte **des Mares**, Croûte **des Blanches Rocques**, Croûte **St Briocq**.
T	Croûte **Bourgeron**, Croûte **Robert**.
F	Croûte **au Blancq**, Croûte **Guille**, Croûte **Norman**, Croûte **Ollivier**, Croûte **Philemon**.
SM	Croûte **des Adoubez**, Croûte **Bertrand**, Croûte **Colin** (**Collas**) **Martin**, Croûte **Corbin**, Croûte **Marinde**, Croûte **Sandre**, Croûte **Thomas Neel** (**Nez**), Croûte **Renault**.
SA	Croûte **Effart**, Croûte **Horaine**, Croûte **Ronchin**.
Cuchon, le	fields: la Mare, SPB. Origin unknown. Recorded **Acuquechon** 1504, **Acuchon** 1663, **Acusson** 1834.
Cucu, la **Pierre** au	land: les Huriaux, SPB. Former l'Erée aerodrome. Surname extinct 15th c. **Le Cucu**.
Cucu, la **Rocque** au	land: les Loriers, CduV. Former common land. Communes q.v.
Cucuel, courtil au	field: les Camps, SM. Surname extinct 17th c. **Le Cucuel** (**Collas Le Cucuel**, 1488).
Cuet, le **Mont**	see **Chouet**, **Crevel**, le **Mont**.
Cuferre, la	field: le Clos Landais, SSV. Origin unknown.
Cuhouel, le **Mont**	see **Chouet**.
Cuissée, la	field: le Hurel, SSV. Origin unknown.
Cure, camp de la	land: S of le Villocq, C.
Cure, la	land, fields, all parishes. Held in the name of the 'living' (la **Cure**) of the parish. Alleur q.v.

D

Dalgairns road	road: between la Gibauderie and Rosaire avenue, SPP. After Captain Dalgairns, a

19th c. British Army officer who owned property in the area prior to the construction of the road.

Dallebec(q) — see **Albecq**.

Damalis, la **Rocque** — land: Icart, SM. Forename (f) **Damalis**. rocque q.v.

Dame, **clos** à la — field: les Bas Courtils, SSV. Presumably reference to ownership in lengthy widowhood.

Dame, de la — fields: N of Fermains, SPP; les Beaucamps, les Pins, C.

Damelaine, de — fields: la Hougue du Pommier, VdelE; les Landelles, C. Surname extinct 15th c. **Dammelaine**. Also feudal holding of land, Fief **Damelaine**, VdelE.

Dames, hougue ès — land: les Pins, C.

Damouettes, des — district: S of Havilland Vale, SPP. Recorded **Danwuette**, 1534. Probably a corruption of Dom Huet (Danhuet).

Damun Gouanne, la **Croix** — site: North side, CduV. Former wayside cross. Recorded 1591. croix q.v.

Dan Hue, la **Capelle** — islet: N of la Croix Martin, SSV. Presumably a hermit's cell. Popularly associated with **Dan Hue** (see above).

Dan Hue, la **Verbeuse** — see **Verbeuse, Dom Hue**. From **Dominus Hue** (surname unknown).

Dan Hue, la **Ville** — see **Ville Dan Hue**. Derivation as above. Also feudal holding of land, Fief du **Domaine Dom Hue**, SSV.

Dan Hue, le **Moulin** — site: Airport, SSV. Medieval windmill, decayed by 17th c. Site since lost. Soil marks destroyed by airport extension, 1960s. Moulin à vent q.v.

Daniel, de — fields: rue des Prés, SPB; rue Perrot, F.

Daniel, la **Vallée Robert** — land: rue des Prés, SPB. Surname extinct 18th c. **Daniel**.

Dan James, de	field: N of les Huriaux, SM. From **Dominus James**, a religious lay brother (surname unknown).
Dan Johan, de	fields: la Houguette, les Caches, SPB; les Laurens, T. From **Dan (Dominus) Jehan**, a religious lay brother (surname unknown). Also spelt **Danjohan**, **D'Anjoint**, **Dan Jouan**.
Dan Nicolle, la **Croix**	see **Revel**, la **Croix Dan** (**Dominus**) **Nicollas**. Also spelt d'**Anne Nicolle**, la **Croix**. From **Dominus Nicolle** a religious lay brother (surname unknown). croix q.v.
Dan Nicolle, la **rue**	road: les Issues (Rocquaine), SPB. Derivation as above.
Dan Philippe, le **Douit**	stream, land: les Landelles, C. From **Dominus Philippe**, a religious lay brother (surname unknown). douit, vau q.v.
Dan Pierre, de	field: near the presbytery, SA. **Dom Pierre**, 1586.
Dans, la **Croûte** ès	field: rue de l'Arquet, SSV. Surname extinct 15th c. **Le Den**.
Dan Thomas, de	field: les Norgiots, SA. From **Dominus Thomas**, a religious lay brother (surname unknown).
Daumale, de	field: les Gigands, SSP. Extant surname now **Domaille**, spelt **Daumale** 14th, 15th & 16th centuries. la Domaillerie q.v.
Dauvel, de	fields: la Mare de Carteret, C. Surname extinct 15th c. **Dauvel**.
Dauvre, le	see **Douvre**.
Davids, les	fields: les Nicolles, SSP; les Annevilles, le Neuf Clos, la Vieille Rue, la Grande Rue, SSV. Extant surname **David**, also spelt **Davy**. Also feudal holding of land, Bordage **Jourdain David**, SSV.
Davy, de **Rosse**	see **Rose**.

De, surnames with an initial De(s) used as proper names for part of a place name, are indexed under the substantive name;

De **Baugy**, De **Beaucamp**, De **Begueville**, De **Bertrand**, De **Beuval**, De **Calais**, De **Carteret**, De **Cobo**, De La **Court**, De **Fosse**, De **France**, De **Garis**, De **Gruchy**, De **Havilland**, Des **Isles**, De **Jersey**, De **Lande**, De **Lestac**, De La **Marche**, De La **Mare**, De La **Mer**, Des **Mares**, De **Nermont** (De Noirmont), De La **Rue**, De L'**Erée** (De Lerée), De **Lespesse**, De **Lisle**, De **Mouilpied**, De **Putron**, De **Quetteville**, De **Saumarez**, De **Sausmarez**, De **Vic**(**q**) (also Devicq).

Debouteurs, les — field: les Fontenelles, SSV. Origin unknown. Spelt **Deboutures** in 19th & 20th centuries.

Dégrais de la **Bouillonne**, les — see **Bouillonne**.

Déhus, le — megalith: N of la Rochelle, Cdu V. existing 'chambered tomb'. Megaliths q.v.

Déhuzet, le — fields: Rozel, SPP; S. of les Eturs, C; l'Erée, SPB. Diminutive of **Déhus**. Minor chambered tomb. None now exist as sites and boulders have totally disappeared. (Pronounced colloquially; 'tuzée'). megaliths, q.v.

Delancey park — area: at Les Monts, SSP. Named after Delancey Barracks. These were named by the Barrack Master General of the British Army, Major General Oliver De Lancy, after his cousin Col. John De Lancy, Asst. Barrack Master in Guernsey at the end of the 18th c.

De le Rée — misspelling of **De Lerée** (Erée, de l', q.v.).

Delisles, les — see **Lisle**, de.

Demaine, du — see **Domaine**.

Demi-Acre, du — field: W of le Grand Fort, SSP. 'half-acre', i.e. 2 vergées Guernsey measure.

Demoiselles, des — field: rue Cohu, C. 'the ladies'.

Denis, de — fields: E of la Longue Rue, SSP; E of le Clos Landais, SSV; la Mare, SPB; Bon Port, SM.

Denis, d'**Abraham**	field: la Croix au Baillif, SA. Surname extinct 17th c. **Denis**.	
Denis, de **Colin**	field: la Haye du Puits, C. Derivation as above.	
Denis, de **Jouanne**	field: Candie, C. Derivation as above.	
Denis Langlois, fosse	land: les Rocquettes, F.	
Denise, de	field: C. Surname extinct 16th c. **Jehan Denise**.	
Dérocquit, sur le	land: 'de-stoned'. SPP. T.	
Derrière, de	field: 'inside' or 'behind'. SSP.	
Désert(s), le(s)	field, all parishes. 'reclaimed waste' (not usually subject to 'champart').	
Désert,	le Carré – square le Grand – big le Long – long le Petit – small	
Désert de la **Glié**	field: S of la Roberge, F.	Glié q.v
Désert des **Brouards**	field: les Merriennes, F.	Brouard q.v.
Désert des **Fourques**	field: N of le Monnaie, SA.	Fourques q.v.
Désert des **Landes Yvelins**	field: les Landes, SPB.	Yvelins q.v.
Désert des **Sommeilleuses**	cliff land: W of Petit Bot, F.	Sommeilleuses q.v.
Désert du **Cardeau**	field: les Vallées, F.	Cardeau q.v.
Désert du **Gal**	field: le Bouillon, SA.	Gal q.v.
Désert du **Trop Loué**	field: Beauvalet, SSV.	Trop Loué q.v.
Désert Gaudin	field: les Pages, SM.	Gaudin q.v.
Désertine, la	land: Grange, SPP; les Pelleys, C; les Fauconnaires, les Rohais, SA. Diminutive of desert. 'reclaimed waste'.	
Des Mares,	see **Mares, Des**.	
Detroit	house: the Grange, SPP. Named by John	

Savery Brock after his brother Major Genl. Sir Isaac Brock's victory at Fort Detroit in 1812. Currently (2012) used by States Education Department.

Devant, de — see **Courtil**.

Devic (**q**), — see **Vic**(**q**).

Deviresse, la — field: les Naftiaux, SA. 'curve, bend.'

Dévise (f) — fields: les Platspieds, C; la Hougue Fouque, SSV; la Dévise, SPB; Rougeval, T; rue des Grons, SM. 'boundary stone or marker'.

Dicqs, les — land: les Sauvagées, les Robergeries, SSP; les Dicqs, VdelE. 'dykes' on low lying land. Also spelt **Dics**, **Dis**, **Dise**, **Ditts**. barrat q.v.

Dicqs, les — road: N of les Dicqs, VdelE.

Dieu le Voye, Fief — feudal holding of land. Dependency of Fief Gohier, SSV.

Dindons, les — field: S of les Videclins, C. Named 1745 as **Gaidons**, (**Guesdons**), q.v.

Distillerie, la — property: les Bergers, C. a liquor 'still', 19th c.

Divette, de — medieval pier, cliff-land: Jerbourg, SM. Origin unknown, possibly a small clear stream. Recorded **Divchete**, **Divete**, 1511.

Dobrée, le — house, field: les Canus, SSP. After **Jean Dobrée** in the 18th c. Surname extinct in Guernsey in late 19th c.

Docques, les — fields: les Martins, SPB. Origin unknown.

Dogon, de — field: N of le Villocq, C.

Dogon, la **Maison** — house, field: Le Préel, C. After **Gilles Dogon**, 1548. Surname extinct 16th c. **Dogon**. Maison q.v.

Dolbel, la **Terre** — land: E of la Ramée, SPP. Surname extinct 19th c. **Dolbel**. terre q.v.

Dolent, le **Camp**	field: CduV. 'flying'. Possibly 'blown away'. Recorded le **Camp Vollent**, 1591. Misspelt in 19th & 20th centuries.
Dom Hue, de	see **Danhue**, **Hue**.
Dom Jean, de	see **Danjohn**.
Domaillerie, la	former house, land: Woodlands, C. After **Jean Domaille**, 1672. Extant surname **Domaille**. Daumale q.v.
Domaine(**s**), le(s)	houses, fields: les Vardes, SSP; les Domaines, le Clos Vivier, SSV. 'domain, estate'. Also feudal holding of land, Fief du **Domaine Dom Hue**, SSV.
Dominique, de	field: le Mont Durand, SPP. Forename (m) **Dominique**.
Dominus Nicolas Revel, le croix	cross: rue des Rocques, T.
Doné, de	field: S of les Baissières, SA. 'given, gift'. Recorded Courtil **Donney**, 1610.
Don Thomas, de	see **Dan Thomas**.
Dorey, de	field: le Friquet, C.
Doreys, la **Hougue** des	houses, land: E of le Rocher, VdelE. After **Jean & Richard Dorey**, 1610. Extant surname **Dorey**. hougue q.v.
Dos d'Ane, le	road, fields: to N and below les Hougues, C. 'the asses' back.' i.e. a rough bumpy roadway.
Douaire, le	field: le Felcomte, SPB. 'dowry', marriage or widow's settlement.
Doublier, la **Coupée**	houses, land: S of rue des Cornets, SPP. Doublier's 'cutting' [lane]. After **Collas Doublier**, 1574. Surname extinct 17th c. **Doublier**.
Douglas, la **Terre**	land: les Rohais, SPP. Surname extinct in Guernsey 16th c. **Douglas**. terre q.v .

Douit, le	houses, fields, all parishes. 'stream'. See Ruel, Ruisseau, Russel.
Douit Benest, rue du	road: unlocated, SPB.
Douit, le camp du	land: near le Bas Marais, SSV.
Douit, le **Grand**	houses, fields: W of la rue Sauvage, SSP; Perelle, SSV. 'large stream'.

Douits, with historical proper names listed by parish. See also individual entries.

SPP	Douit des **Longs Camps**, Douit du **Foulon**, Douit **de la Platte Hache**, Douit **De Havilland**, Douit **des Moulins de la Vrangue**, Douit **de la Planque**, Douit du **Pont Renier**, Douit **De Putron**, Douit des **Rohais**.
SSP	Douit **des Orgeris** or Douit **de la Tonnelle**.
VdelE	Douit **Brazar**, Douit **du Bullaban**, Douit **du Pont Perrin**.
C	Douit **des Bergers**, Douit **Brazar**, Douit **des Fauquets**, Douit **du Boudin**, Douit **du Moulin**, Douit **du Grand Moulin**, Douit **de la Pitouillette**, Douit **Pelley**, Douit **Dan Philippe**, Douit **du Villocq**.
SSV	Douit **des Choffins**, Douit **Manchot**, Douit **du Moulin**.
SPB	Douit **Benest (Benoist)**, Douit **de Beuval**, Douit **d'Israel (Le Lacheur)**, Douit **du Moulin**.
T	none.
F	Douit **des Landes de la Foret**.
SM	Douit **de la Bette**, Douit **des Ménages**, Douit **de la Porte**, Douit **ès Tardifs**.
SA	Douit **Marin**.

Douit, de **Jean du**	field: l'Alleur, T. Surname extinct 17th c. **Du Douit**. Also feudal holding of land, Bouvée **Richard Du Douit**, C.
Douvre, le	houses, land, fields, all parishes. 'embankment'. Also spelt **Dauvre**.
Douvre Colin Gringoire, le	shelter, embankment: le Felcomte, SPB. See Gringoire.

Douvre, le camp du	land: Bon Port, SM.
Douvriers, les	land: le Crolier, T. derived from douvre, q.v.
Doyle column	monument: le Mur, Jerbourg, SM. After Major General Sir John Doyle, Lieutenant Governor, 1803-16. Rebuilt 1953 following destruction during the German Occupation.
Doyle, **Fort**	fortification: E end of Fontenelle Bay, CduV.
Doyle road, street	roads: Grange, le Truchot. SPP. After Sir John Doyle, as above.
Doyle, rue	road: in Braye du Valle, CduV. Derivation as above. Now called Route Militaire, q.v.
Doyles, les	land: l'Ancresse common. Mid-19th c. barracks and dwellings. Derivation as above.
Driet, de	field: N of la Turquie, CduV. Origin unknown. Recorded 1591.
Drognet, de **Marquet**	field: Woodlands, C. Surname extinct 1470. **Drognet**.
Drouins, les	field: la Pomare, SPB. Surname extinct 15th c. **Drouin**.
Drouins, clos	field: now in reservoir, SSV. Derivation as above. clos q.v. Also feudal holding of land **Boissel du clos Drouin**. VdelE.
Dry, la **Hure**	land: la Hougue Taive, T. After **Michel Dry**. Surname extinct 15th c. Druy, Dry, camp, hure, Mondril (Montdril) q.v.
Dry, le **Camp**	land: les Landes, CduV.
Du	surnames with an initial **Du**, used as proper names for part of a place name, are indexed, with a few exceptions, under the substantive name; Du **Douit**, Du **Four**, Du **Frocq** (Fro), Du **Gaillard**, Du **Genât**, Du **Hurel**, Du **Maresq** (Dumaresq), Du **Mont** (**Dumont**), Du **Paysan**, Du **Port** (Duport), Du **Quemin** (Duquemin), Du **Val**, Du **Vauriouf**.

Dugnet, la maison — land: les Vesquesses, C.

Dumaresq, le **Clos** — fields: rue des Bergers, C. Extant surname **Dumaresq**.

Dune, la — fields: les Fauquets, C. Surname extinct 17th c. **Dune**.

Dunes, les — sand dunes, sandy fields: l'Ancresse, CduV; Portinfer, VdelE; Vazon, C.

Duquemin, le **Clos** — land: les Grands Moulins, C. Extant surname **Duquemin**. Also feudal holding of land, la Bouvée **Duquemin**, SPB.

Durand, la **Vallée** — houses: les Petites Fontaines, SPP. Also feudal holdings of land, Bordage **Durant**, SPP and Fief **Durant**, SM.

Durant, le **Mont** — houses, land: le Mont Durand, SPP; le Mont Durand, SM. Surname extinct in Guernsey 20th c. **Durand**.

Durel, de — land, fields: les Carterets, C. Surname extinct 15th c. **Durel**.

Duriaux, la **Ville** es — land: S of les Damouettes, SPP. House demolished mid-19th c. Plural of above extinct surname **Durel**. Also feudal holding of land, Bouvée **Durel**, C.

Du Val, le **Courtil** — field: N of des Duvaux, SSP. Surname extinct 19th c. **Du Val** (extant Jersey). Misspelt in plural as **Duveaux**. val q.v.

Duvaux, les — houses, land, fields: E of Baubigny, SSP.

E

Eau, la **Rocque** à l' — land: les Choffins, C. 'rock by the water' (i.e. douit).

Eau, la **rue** à l' — road: le Torval, C. 'water lane'.

Ebat, le **Camp** d' — fields: W of Rougeval, T. Origin unknown. camp q.v.

Ebénézer, d’	building, house: Sausmarez street, SPP; house W of la Pomare, SPB (now renamed). Former chapel and house name, 19th c.
Ecachier, Fief de l’	feudal holdings of land. There are two small fiefs with this name, not apparently connected. One is a dependency of Fief de Carteret, C; and the other of Fief des Vingt Bouvées du Villain Fief le Comte, C.
Ecanges, les	see **Equanges**.
Echèlle, de l’	house, water mill: W of les Galliennes, SA. upper and lower mill, called **Scala** or **Escala**, 1331. One mill still exists. Possible connection with ‘ladder’. moulin à eau q.v.
Echèlle, rue de l’	road: unlocated, SA.
Echèlons, les	house, roadway: South Esplanade, SPP. ‘steps’, en echelon. 19th c.
Eclet, l’	house, fields: N of la Couture, SPB. Origin unknown. Misspelling of **Equelley**, 19th & 20th c.
Eclet, rue de l’	road: N of route des Paysans, SPB.
Ecluse, l’	fields: Sous l’Eglise, SSV; le Vallon SM. Mill pond, to feed a water mill. moulin à eau q.v.

Ecluses (unnamed) with approximate locations, listed by parish:

SPP	le Bordage, below la Charroterie, below le Pont Renier, la Vrangue.
CduV	none.
VdelE	none.
C	le Grand Moulin (Water Board), E of le Moulin du Milieu, above le Moulin de Haut.
SSV	at ruette de la Potère (Beauvalet), above Moulin de la Perelle (now in Reservoir).
SPB	above Moulin de Becquet (Water Board), above Moulin de Cantereine.
T	none.

F	les Glayeuls and E of Le Variouf, rue du Forval (Petit Bot).
SM	above Moulin Huet.
SA	above Moulin des Niaux.
Ecluse Corbin, l’	mill pond: les Poidevins, SA. Extant surname **Corbin**.
Ecluse du **Moulin Foullères**, l’	pond: le Vauquiedor, SA. Moulin à eau q.v. mill
Ecole, **Croûte** de l’	field: le Tertre, CduV. Origin uncertain. croûte q.v.
Ecole du **Bosq**, l’	site: Glategny (N of la Petite Ecole), SPP. Reference 1574.
Ecole Publique, l’	site: between les Prés and la Vallée, SPB. Demolished mid-18th c and replaced by school at les Buttes, SPB.
Ecole, la **Petite**	site: bottom of St Julian’s avenue, SPP. Former late medieval Town school, demolished 19th. Adjacent to former Chapelle St Julien q.v.
Ecole, la **Vieille**	old school: Castel School, C. Site of la masse (mound) on which was built le moulin à vent (windmill) of Fief le Comte. masse, moulin à vent q.v.
Ecole, la **Vieille**	old school: St Martin’s school, SM. Site of former Chapelle de St.Jean de la Houguette q.v.
Ecollasse, la **Croûte**	field: la Couture, SPP. Surname extinct 16th c. **Escollasse**.
Ecollasse, la **Saudrée**	field: Les Baissières, SA. saudrée q.v.
Ecrilleurs, les	cliffs: T. Origin unknown. Possibly from escrillier, to slip. Spelt **Ecrillier**, 1705.
Effard, les	houses, fields: W of Baubigny, St Clair, SSP; E of les Beaucamps, W of la Hougue Renouf, C; le Carrefour au Lièvre, la Fallaize, SM; la Contrée de l’Eglise, SA.

Effards, la **Croix** des — site: les Effards, SSP. Former wayside cross (unlocated). Common surname extinct 19th c. **Effard**, also spelt **Effart**. croix q.v. Also feudal holding of land, Fief des **Effards**, C.

Effards, rue des — road: N of rue du Préel, C. Now called Effards Lane.

Effart, croûte — land: unlocated. SM.

Eglise, **Courtil** de l' — land: les Rohais, SPP; le Vauquiédor, SA. Former fields, previously owing rentes to the parish church.

Eglise, la **Contrée** de l' — houses, fields: near parish church, SA. 'district of the church'.

Eglise, le **Chemin** de l' — roads, ways, all parishes. Direct 'way' along roads and lanes from the extremities of each parish to the parish church, for services and funerals. Also called **Rue de l'Eglise**.

Eglise, **Sous** l' — houses, fields: N and below parish church C; valley below parish church,SSV. 'below the church'.

Eglises Paroissiales, les — listed by parish.

SPP	**St Pierre Port**
SSP	**St Samson**
CduV }	**St Michel du Valle**
VdelE }	
C	**Ste Marie du Castel** (previously **Notre Dame du Castel**)
SSV	**St Sauveur**
SPB	**St Pierre du Bois**
T	**St Philippe** (earlier church **Notre Dame du Torteval**)
F	**Ste Marguerite** (previously la **Trinité de la Foret**)
SM	**St Martin de la Bellieuse**
SA	**St André** (de putenti pomerio)

Elaine, d' — field: l'Aumone, C. Forename (f) **Elaine**. Reference 18th c.

Elizabeth Gauvain, d’	see **Gauvain**.
Elizabeth Rabey, de	see **Rabey**.
Elliot, d’	field: les Martins, SPB. Origin unknown, first recorded 1700.
Emerais, les	house, fields: E of les Houmets,C. After **Mychiel Emerey**, 1595.
Emerais, ruette des	road: N of Saumarez Park, C.
Emon, d’	field: la Fallaize, SM. Forename (m) **Emon**. Also feudal holding, le Bordage **d’Emon De La Rue**, T.
Emrais, la **Croûte** des	land: les Houmets, C. Surname extinct 17th c. **Emerey**.
Enclos (m)	curtilage or immediate land holding of a house and fields.
Enfant, de **Bon**	see **Bon Enfant**.
Enfants, les	field: les Houets, T. Presumably inherited and owned for a long time by minors.
Enfer, l’	fields: S of la Grande Mare, C; le Hurel, SM. ‘hell’, i.e. appalling, terrible.
Enfer, le **Clos** d’	field: les Vallettes, SSV. Derivation as above. clos q.v.
Enfer, le **Port**	bay: E of les Grandes Rocques, C. Derivation as above. port q.v.
Enfer, rue d’	road: N of King’s Mills, C.
Eperons, les	house, fields: les Eperons, SA. Also feudal holding of land. Fief des **Eperons** SA. Shown as two separate holdings in Extente of 1331, each owing pair of gilt spurs to Crown annually. Once amalgamated, only owed one pair. Despite above, there is no evidence of any such payment.
Eperquerie, l’	land: le Pezeril, CduV; les Pezerils, Pleinmont, T; Saints, SM; Jerbourg, SM. ‘drying place for conger eel and mackerel’. Pezeril q.v.

Epine Gaudin, l'	see **Gaudin**.
Epine Touzee, l'	field: les Camps, SM. Origin uncertain, first recorded 1768. Possibly from extinct surname **Touzel**.
Epine(s), le(s)	land, fields: N of la Vrangue, SPP; N of Delancey, SSP; le Tertre, la Rochelle, les Petils, CduV; le Douit, C; la Fallaize, les Martins, Saints, le Mont Durand, SM. 'thorn', i.e. former wasteland.
Epine, la **Chapelle** de l'	site: le Hamel, VdelE. Former chapel of ease to Vale Church, with cemetery (abandoned and ruined before the Reformation). Dedicated as Notre Dame de l'Epine.
Epine, la **Vingtaine** de l'	area of parish du Valle outside le Clos du Valle. Derivation as above. chapelle, val, Vingtaine q.v.
Epinel, l'	fields: S of the Braye du Valle, SSP; la Croix Foret, F. Diminutive of epine, 'thorn'.
Epinelle, l'	fields: E of les Gigands, SSP. Derivation as above.
Epines, le **Val** des	land: Pleinmont, T. Derivation as above, 'thorn'.
Epinette Mahaut, l'	land: N of les Rohais, SPP. Diminutive of epine, 'thorn'. Mahaut q.v.
Epinette, l'	fields, all parishes. Derivation as above.
Epis, de l'	field: le Lorier, SSV. Origin unknown, recorded **Epy**, 1723.
Epointal, l'	land: W of le Gouffre, F . 'blunted point'.
Equanges, les	field: Jerbourg, SM. Origin unknown. See also Ecanges.
Equelley, l'	see **Eclet**.
Equête, l'	field: la Hougue, C. Origin unknown. Possibly originally spelt l'enquête.
Erclin Rocque, l'	see **Rocque**, **Roquelin**.

Erée, l'	promontory: facing Lihou Island, SPB.
Erée, la **Croix** de l'	site: l'Erée (on ridge), SPB. Former wayside cross. Croix q.v.
Erée, la **Tour** de l'	fort: Martello tower, q.v. Fort Saumarez q.v. Surname extinct 20th c. **De Lerée**. Misspelt **De Le Rée** 20th c. Also feudal holding of land, Fief des **Trois Bouvées de l'Erée**, SPB, which comprises the whole promontory.
Erée, les **Casernes** de l'	buildings: below Fort Saumarez, SPB. Former barracks. Casernes q.v.
Eri, la **Hougue**	land: les Grantez, C. Presumably extinct surname. Recorded **Hougue Erreye**, 1512.
Escaliers, les	land, fields, all parishes. Possibly 'stepped'.
Escalliers, rue des	road: E of la Villette, SM.
Escollasse, croute	land: near la Couture, SPP.
Espesse, Fief de l'	feudal holding of land. Subsequently known as Fief au **Canely**, which was divided between various heirs in the early 13th c. T.
Esplanade, les **Valettes**	roadway: on S of Havelet bay.
Esplanade, **North**	roadway: alongside Careening Hard, SPP harbour.
Esplanade, **South**	roadway: alongside Havelet bay. Le Galet Heaume q.v.
Esplanade, **St George's**	roadway: alongside La Salerie harbour.
Esquière, la **Forfaiture** a l'	land: Les Forfaitures, C.
Esquièrre, la **Fosse** à l'	house, land: la Villiaze, SA. Forfaiture of **John Lesquyerre**. From **John Squires**, an Englishman, 16th c.
Essart(s), le(s)	land: Baubigny, le Marais, SSP; Sohier, CduV; la Marette, rue de l'Arquet, SSV.
Essart, la **Fosse** de l'	land: rue de l'Arquet, SPB. 'assarted land'. Clearance of waste outside the then

	accepted arable into strip cultivation.
Essel, fontaine	spring: rue a l'Or, SSV.
Essel, l'	land: les Grand Moulins, C; rue à l'Or, SSV; les Cornus, SM. Origin unknown. Also spelt **Essay**.
Essilleril, l'	fields: les Simons, T. Possibly 'deep furrow'. Recorded **Essylleryll**, 1592. Also spelt **Esillery**. Misspelt **Sillery**, 20th c.
Esther, d'	field: les Poidevins, SA. Forename (f) **Esther**.
Estrainfer, l'	coastal area: Côbo, C. 'rough wild beach'.
Etac, (**Etat**), de l'	fields: rue à l'Or, SPP; les Monts Hérault, SPB; le Crôlier, T; la Pompe, SM. Surname extinct 16th c. **De Lestac**.
Etac, le **Clos** de l'	field: les Rébouquets, SM. Derivation as above.
Etang, **Etant**, le **Mal**	field: les Choffins, SSV. 'bad (ororous) pond'.
Etat, le **Hure** de l'	field: S of le Crôlier, T. Derivation as above
Etibots, les	fields: les Baissières, SPP, VdelE. Origin uncertain.
Etienne, d'	fields: les Francais, C; N of les Quevillettes, les Pages, SM. Late of **Etienne Martin**, 16th c.
Etienne, de **Colin**	field: CduV. Late of **Colin Etienne**, 1654.
Etiennerie, l'	house: N of Les Jaonnets, C. built by **Etienne Gavet**, 1730/50.
Etonnellerie, l'	house, land: N of rue Maingy, VdelE. Surname extinct early 19th c. **Lestournel**. Misspelling of **Lestournelerie** by OS surveyors, 1898-1900. Etournel q.v.
Etoquet, l'	fields: Chouet, CduV. Origin unknown. Recorded **Letoquet**, 1591.
Etots, les	field: les Fauquets, C. Origin unknown

Etourneaux, l'	former house: les Cherfs, C. From extinct surname **Lestournel**. Also spelt **Etournios**. See **Etonnellerie**.
Etournel, l'	fields: les Vardes, SSP; la Lande, C. Misspelt **Le Tournel** 18th c.
Etur, de **Simon**	land: W of les Niaux, C. Surname extinct 17th c. **Etur**.
Etur, la rue	road: now called rue des Goddards, C. Derivation as above.
Etur, le **Val**	fields: la Croisée W of les Tielles, T. Derivation as above.
Eturs, les	district, fields: les Baissières, N-E of les Grands Moulins C. Derivation as above.
Eveque, Fief de l'	feudal holding of land. Originally held by the Bishop of Coutances. Now a Crown fief. SA.

F

Fabien, le **Clos**	land: unlocated, SPP. late of **Nicolas Fabien**, 1658.
Fainel, le	house: le Hurel, SM. 19th c. house name.
Fainel, ruette du	lane: le Fainel, SM.
Fairfield	field: N of route de l'Eglise, C. Used for 19th c. cattle shows.
Falla, de	fields: les Marais, la Ronde Cheminée, Pied des Monts, SSP; (unlocated),CduV. Extant surname, **Falla**.
Fallaize, **de**	fields: les Mares, SPB; la Quevillette, SM; les Norgiots, SA. Extant surname **Fallaize**.
Fallaize, de **Flocel**	field: les Maumarquis, SA. After **Flocel Fallaize**, 16th c. Guernsey merchant. Derivation as above.
Fallaize, fontaine	spring: near le Vallon, SM.

Fallaize, la	houses, land, fields: Havelet, SPP; rue de la **Fallaize**, SM. 'cliff'. From surname, Pierre **Fallaize**. Also spelt **Falaise**. Also feudal holding of land, Fief **de la Veleresse du Côté de la Fallaize**, SM, q.v.
Fallaize, la **Croix**	field: les Coutures, SM. wayside cross. Derivation as above. croix q.v.
Fallaize, la **Rouge**	field: les Petils, CduV. 'reddish cliff'. Derivation as above.
Fallaize, le **Coin**	road junction: S-E of les Blanches, route de Jerbourg, SM. Extant surname **Fallaize**.
Fallaize, rue de la	road: now called Cliff St, SPP; E of Petit Bot Bay, SM.
Fallaizes, les **Courtes**	house, field: le Hurel, SM. Named after **Jean** and **Thomas Fallaize**, 18th c. messières, q.v.
Fallaizettes, les	slopes: le Moulin de haut, C. Diminutive of **Fallaize**. 'low cliffs'.
Falles, les	fields: E of les Sages,SPB. Surname extinct 15th c. **Falle** (extant Jersey).
Falle, le **Hougue**	fields: W of les Vinaires, SPB. Derivation as above. hougue q.v.
Fallue, la	land: les Godeines, SPP. Possibly a 'slope'. Tallut q.v.
Fanigot, de	fields: les Beaucamps, C; W of les Annevilles, SSV. Surname extinct 16th c. **Fanigot**.
Fantôme, le **Bordage**	feudal holding of land, unlocated. SSP. Bordage q.v.
Faras, **Farras**, le	district: W of les Landes, F. Recorded 'contrée de **Lengage Farras**'. 1504. Possible meaning 'Fara's pledge'.
Fauconnaires, la **Croix** des	site: unlocated, les Fauconnaires, SA. wayside cross. Recorded 1610 (**Pierres le Fauconnere**, HM Receiver, 1492.) Surname extinct 16th c. **Le Fauconnaire**.

Fauconnaires, les	district: S of les Rohais, SA.
Faussillon, le	field: le Carrefour, C. Surname extinct 14th c. **Faussillon**.
Fautrart, le	fields: Choisi, SPP; la Hougue Fouque, SSV; les Blanches Pierres, SM. Surname extinct 17th c. **Fautrart**. Misspelt **Fautrac**, 19th & 20th centuries. Cardin q.v.
Fauvel, le	house, fields: les Prevosts, SSV. Surname extinct 16th c. **Fauvel** (extant Jersey).
Fauville, Fief de	feudal holding of land. Dependency of Fief d'Anneville, SSP.
Fauxquets, douit des	stream: Fauxquets Valley, C.
Fauxquets, les	district: W of Candie, C. Surname extinct 15th c. **Fauque**. Also spelt **Fauquets**, **Fauquez**.
Favière, la	field: le Bas Marais, SSV.
Faviole, la	land: Hacsé, CduV. Recorded **Faviolle**, 1564. Possibly from taviole, 'small level plot'.
Fayes, le **Fossé** ès	land: Pleinmont, T. 'fairies' bank'.
Febvre, la rue **Le**	see **Lefebvre**, la rue.
Febvre, le	field: les Fosses, F.
Febvre, le **Val** au	fields: S of les Blicqs, SM.
Fee farms, H.M.	
SSP	**la Thée**
CduV	**My Lady** (le Rond Marais)
C	**la Grande Mare, la Mare Dalbec**
SSV/SPB	**la Claire Mare**
SPB	**la Rousse Mare**
SA	**les Vauxbelets**
–	**Lihou Island**
	Fee farms are Crown land which has been let on long leases since 1737.
Fées, camp des	land: Bon Port, SM.

Fées, la **Rocque** aux	land: le Bourg, F. 'fairies' rock'.
Fées, le **Creux** ès	megalithic tomb: l'Erée, from creux, 'fairies' hollow'.
Feivre, au	see **Feyvre**.
Felcomte, la route du	road: S-W of les Rouvets, SSV, SPB. 20th c. corruption of **'Fief le Comte'**. Misspelt Felconte. Recorded la **rue de la Bette**, 1633. Bette q.v.
Femme, de la **Bonne**	field: les Blicqs, SA. Possibly from rente owed to a medieval church confraternity for women.
Fenêtres, les	fields: les Houards, F. Origin uncertain.
Ferée, la rue	roads: W of King's road, Mount row, la Vrangue, SPP; la Mare de Carteret, W of la Hougue, C. 'metalled' road.
Ferenbras, de	field: la Haye du Puits, C. After **Ferenbras le Jersiez**, 1548.
Fermain(s), de	port, bay, guet: S of Fort George, SPP. field; Calais, SM. Possibly from an unrecorded surname, **Fermains**. Also feudal holding of land, Fief **de la Veleresse du Côté de Fermains**, SM. port, guet q.v.
Fermains, ruette sur	lane: E of Fort Road, SPP; la Bouvée, SM.
Fermes, ès	field: le Hurel, CduV. Origin unknown. Recorded 1654.
Ferre, rue du	road: unlocated, C.
Ferron, la **Vallée** au	field: les Hougues, C. Surname extinct 16th c. **Ferron**. vallée q.v.
Fers à **Chanvres**, les	field: la Hure Mare, CduV. 'hemp store'.
Feugère, la	field: les Landes Hues, T. 'bracken'.
Feugré, **Feugrel**, le	fields: all parishes. Contraction of feuguerel, 'bracken brake'.
Feugré, le	road: W of ruette de la Tour, C.
Feugriaux, les	fields, all parishes. Plural of **Feugrel**. Also

	spelt **Feugriots**, **Fueguerriaux**. Derivation as above.
Feveresse, rue	house, field: W of les Picques, SSV. Origin unknown. 1st ref. 1853.
Féves, à	field: les Villets, F. 'beans'.
Feyvre, au	field: la Lande, C.
Feyvre, la rue au	road: N of les Grands Moulins, le Pavé, C. Called rue des Belles since late 20th c. Extant surname **Le Feuvre**, from Le **Feivre**, 18th c. Febvre, Le q.v. Also feudal holding of land, Fief des **Feyvres**, C.
Fiacre, le	field: le Soucique, F. Forename (m) after **Fiacre De Lerée**, mid-16th c.
Fiaubré, le	field: N of Bon Port, SM. From Fief **Aubret**. Brayes, Brets q.v.
Ficquet, le	field: les Fourques, SA. Late of **Collas Ficquet**, 1611.

Fiefs, feudal holdings of land held by seigneur or dame (or the Crown) listed by parish. See also individual entries.

SPP	Fief **De Rozel**, Franc Fief **Saint Martin**, Fief **le Roi**.
SSP	Fief **D'Anneville**, Fief **des Bruniaux de Nermont**, Fief **de Fauville**, Franc Fief **Gallicien**, Fief **Henry de Vaugrat**, Fief **des Philippes**, Fief **le Roi**, Franc Fief **De La Rosière**.
CduV	Fief **Saint Michel**.
VdelE	Fief **Bournel**, Fief **du Camp des Haies**, Fief **au Carpentier**, Fief **au Legat**, Fief **du Quartier du Camp Rouf**, Fief **du Quartier des Goubeys**, Fief **Richard De La Folie**, Fief **Richard De Nermont**, Fief **Roger Gosselin**, Fief **De Rozel**, Fief **Saint Michel**, Fief **de la Vingt-et-unième Boisselée**.
C	Fief **de la Belenger**, Fief **des Besoignes**, Fief **de la Bouvée du Groin**, Fief **Bertram**, Fief **au Breton**, Fief **de la Cannevière**, Fief **De Carteret**, Fief **de la Chapelle Saint George**, Fief **des Cherfs**, Fief **au Chevalier**,

	Fief **des Cobois**, Fief **au Cocq**, Fief **Cohu**, Fief **le Comte**, Fief **des Corvées**, Fief **De La Cour**, Fief **Covin**, Fief **Dom Jean le Moigne**, Fief **a l'Ecachier** (1), Fief **a l'Ecachier** (2), Fief **des Effards**, Fief **des Feyvres**, Fief **des Frohiers**, Fief **des Grangiers**, Fief **des Grantez**, Fief **du Groignet**, Fief **des Hasios**, Fief **Herne**, Fief **de la Landelle**, Fief **Lucas Arnault**, Fief **des Moullinez**, Fief **du Naunage**, Fief **ès Pellais**, Fief **des Prestres**, Fief **des Queus**, Fief **ès Riollais**, Fief **de la Rivière**, Fief **des Rompeurs**, Fief **au Rougier**, Fief **Saint Michel**, Fief **De Saumarez**,Fief **au Saunier**, Fief **Sotuas**.
SSV	Fief **de la Boisselée de Henry du Vauriouf**, Fief **de la Bouvée Marquand**, Fief **de Cantereine**, Fief **le Comte**, Fief **des Dix Quartiers Blondel**, Fief **Dieu le Voye**, Fief du **Domaine Dom Hue**, Fief **des Fouquées**, Fief **des Gervaises**, Fief **des Gohiers**, Fief **Gouie**, Fief **Hillaire**, Fief **Huchon**, Fief **Jean Du Gaillard**, Fief **de Lihou**, Fief **de Longues**, Fief **des Mauconvenants**, Franc Fief **des Reveaux**, Fief **au Roux**, Fief **Saint Michel**, Fief **de Suart**, Fief **des Trois Vatiaux**.
SPB	Fief **Bequepée**, Fief **De Beuval (formerly Fief de Lespesse**), Fief **le Comte**, Fief **ès Corbinez**, Fief **de la Bouvée Duquemin**, Fief **de la Cour Ricart**, Fief **au Crochon**, Fief **Janin Bernard**, Fief **Jean du Gaillard**, Fief **de Lihou**, Fief **au Mière**, Fief **Pomare (d'Appemare**), Fief **des Reveaux**, Fief **Robert de Vicq**, Fief **Saint Michel**, Fief **Suart**, Fief **Thomas Blondel**, Fief **des Trois Bouvées De Lerée**.
T	Fief **D'Allebecq**, Fief **des Coltons**, Fief **à l'Eperon**, Fief **Guillot Justice**, Fief **des Huit Bouvées**, Fief **Janin Bernard**, Fief **Jean Du Gaillard**, Fief **de Lihou**, Fief **de Pleinmont (Quarante Quartiers**), Fief **Robert De Vicq**, Fief **Saint Michel**.
F	Fief **le Roi**.
SM	Fief **Avallet**, Fief **de Beauvoir**, Fief **de Blanchelande**, Fief **des Bruniaux**, Fief

	Capis, Fief **Durant**, Fief **Fortescu**, Fief **Geffrey De Begueville**, Fief **Hailla**, Fief **Henry de Castel**, Fief **Lesant**, Fief **Levin**, Fief **au Marchant**, Fief **Massy Gros**, Fief **Richard De Camps**, Fief **Rogier le Villain**, Fief **le Roi**, Franc Fief **Saint Martin**, Fief **De Sausmarez**, Fief **de la Veleresse du Côté de la Fallaize**, Fief **de la Veleresse du Côté de Fermains**.
SA	Fief **de l'Abbesse de Caen**, Fief **des Eperons**, Fief **l'Eveque**, Fief **De La Haule**, Fief **des Maumarquis**, Fief **de la Rue Frairie**, Fief **des Rohais**, Fief **le Roi**, Fief **Sainte Hélène**.
Fief **Herne**, le	field: la Rivière, C. On **Fief Herne**.
Fief le **Comte**, le	fields: les Rouvets, SPB/SSV. On Fief le Comte. Also feudal holding of land, Fief **le Comte**, C/SSV/SPB. Name from the Earls of Chester, the original seigneurs.
Fief le **Comte**, route du	see **Felcomte**.
Fief, le **Franc**	fields, road: N of la Robergerie, SSP. After **Franc Fief Gallicien** in that district (a franc fief being free of most feudal dues.)
Fief **l'Eveque**, le	field: les Gouies, SA. On **Fief l'Eveque**.
Fief **le Roi**, le	1. field: le Neuf Chemin, SSV. On 'king's fief' (**Fief Suart**). 2. house: les Varendes, SPP. On **Fief le Roi** SPP.
Fief Suart, camps du	land: rue de l'Arquet, SPB. 'strips' of land on Fief **Suart**. camp q.v.
Fier Mouton, le	houses, field: S of les Truchots, SA. Origin uncertain.
Fiffaigne, de	see **Tiffaigne**.
Figuier, le	fields: les Petites Vallées, C; les Rohais, SA. 'fig tree'.
Fil, le **Grand**	fields: Mont Plaisant, C. 'big fief'. Areas of strips of land in large fields, mostly on **Fief le Comte**.

Fillage, le — fields: S of la Grande Rue, SSV. Possibly a rope walk.

Fillage, rue du — road: unlocated, SSV.

Filles, ès **Belles** — see **Belles Filles**.

Fiott, pré — field: les Osmonds, SSP. Surname extinct 19th c. **Fiott**. Misspelt **Frott** 20th c.

Flaguée, la — fields: les Clos Landais, SSV. 'levelled land'. Adjacent to (levelled) Hougue Maban q.v.

Flère, de — fields, road: le Maresquet, CduV. Surname extinct 19th c. **Flère**.

Fleuri(**e**), de — fields: les Landelles C; le Rocher, T.

Fleurie Dalbec, la **Fontaine** — land: Havelet (Fontaine Fleury), SPP. Albecq, q.v.

Fleurie Hailla, de — land: les ruettes Brayes, SPP. Forename (m) **Fleurie**. Albecq, fontaine, Hailla q.v.

Fleurie Hailla, ruette — lane: unlocated. SPP.

Foin, clos au — field: les Maumarquis, SA. 'hay'. clos q.v.

Foire, la — field: N of les Touillets, C. 19th c 'fair ground' for cattle shows by Royal Guernsey Agricultural & Horticultural Society. Fairfield q.v.

Folie, la — fields: l'Islet, SSP; S of Ville Baudu, CduV; Pleinmont, T. Surname extinct 15th c. **De La Folie**. Also feudal holding of land, Fief **Richard De La Folie**, VdelE.

Follet, la **Rocque** — land: la Rocque, la Hougue Anthan, SPB. Derivation as above. rocque q.v

Follet, le **Camp** — land: Saints, SM. Surname extinct 15th c. **La Folley** (Laffoley extant Jersey). camp q.v.

Fonderille, la — fields: rue de l'Arquet, SPB; E of les Tielles, T. 'rill' or 'runnel' on spring line. bouillon, fontaine q.v.

Fonds de la **Bourse**, le — land: S of le Villocq, C. 'bottom of the bag'. (colloquially 'small change') i.e. of little appeal, as the land is on the spring-line.

Fontaine ès **Corbins**, ruette de la	lane: S of les Canichers, SPP.
Fontaine, croûte de la	land: near St Clair, SSP.
Fontaine(**s**), la (les)	houses, fields, all parishes. 'springs'. bouillon, fonderille q.v.
Fontaine, la **Sèche**	field: Pleinmont, T. 'dry spring'.
Fontaine, rue de la	road: Fountain street, SPP.

Fontaine(**s**) with proper names, listed by parish. See also individual entries.

SPP	Fontaine **Bouillante de la Couture**, Fontaine **de la Cache Vassal**, Fontaine **ès Corbins**, Fontaine **de Nôtre Dame au Mont Gibet**, Fontaine **Fleurie** (**Dalbec**), Fontaine **Foullo**t, Fontaine **des Grons**, Fontaine **ès Gros**, Fontaine **Tourgis**, Fontaine **de la Ville ès Duriaux**.
SSP	Fontaine **Angot**, Fontaine **St Clair**.
CduV	Fontaine **Perrottes**.
VdelE	Fontaine **du Gal**, Fontaine **Pelichon**.
C	Fontaine **des Boulains**, Fontaine **au Bourg**, Fontaine **au Brun**, Fontaine **des Gaidons** (**des Dindons**), Fontaine **Gonebec**, Fontaine **Guillaume le Mière**, Fontaine **du Puits De Caubo**, Fontaine **St George**, Fontaine **Saint Germain**.
SSV	Fontaine **dit l'Assaut**, Fontaine **au Courly**, Fontaine **des Bas Courtils**, Fontaine **de l'Essel**.
SPB	Fontaine **au Barsier**.
T	Fontaine **Bellée**, Fontaine **de Rougeval**, Fontaine **du Val Aubert**.
F	Fontaine **des Nouettes**.
SM	Fontaine **de la Bette**, Fontaine **ès Colons**, Fontaine **De La Court**, Fontaine **ès Cotentins**, Fontaine **Fallaize**, Fontaine **Jean Fallaize**, Fontaine **Hamelin**, Fontaine **du Hurel**, Fontaines **des Ménages**, Fontaine **des Nicolles**, Fontaine **Rauf**, Fontaine **Ricart**, Fontaine **Robin**, Fontaine **de la Roquaze**, Fontaine **Ros**, Fontaine **Saint**

	Martin, Fontaine **du Varclin**, Fontaine **de la Bouvée**.
SA	Fontaine de **Jehan Denis**.
Fontaines, les **Petites**	road: N of Mont Durand, roadway to Dégrez de la Bouillonne, SPP. 'small springs'. bouillonne q.v.
Fontenelle(**s**), la (les)	houses, fields, all parishes. Diminutive of fontaine, 'small springs'. Coudré q.v.
Fontenelle au **Coudré**, la	land: le Coudré, SPB. Derivation as above.
Fontenil, le	fields: S of les Reveaux, SPB. Variation of fontenelle.
Forêt, la **Croix**	site: E of l'Epinel, F. Former wayside cross. croix, rue q.v.
Forêt, la **rue**	houses, road: Forest lane, SPP. Surname extinct 15th c. **Forest**.
Forêt, le **Douit** des **Landes** de la	stream: Farras, F. (parish boundary). douit, landes q.v.
Forfaiture(**s**), la (les)	field: Sausmarez, SM. Origin unknown. Possibly forfeiture through non-payment of rentes, etc. Esquière q.v.
Forfaiture, camp de la	land: off rue du Felcomte, SPB
Forge(**s**), la (les)	1. houses, road: rue des Forges (Smith street) SPP. 2. houses: rue du Pré, Longstore, SPP; rue Maingy, VdelE; Grands Moulins, C; les Brehauts, les Marchez, SPB; le Belle, T; le Bouillon, SA. Forges, smithies, 19th c. 3. fields: les Issues, SSV; la Houguette, les Buttes, SPB; les Norgiots, SA. 'historical' field forges, smithies.
Forge, camp de la	land: la Houguette, SPB.
Forge au **Feyvre**, la	land: Jerbourg, SM. 'Le Feyvre's forge'.
Forges Gallie, la	field: unlocated, SA. 'Gallie's forge'.
Forgettes, les	houses, field: les Forgettes, C. Diminutive of forge.

Forpas, le — house, land: le Forpas, CduV. Origin unknown.

Fort, le **Grand**
1. large retaining embankment to former saltpans (salines) below les Basses Capelles on W., at W end of former Braye du Valle, SSP. Probably medieval.
2. houses, fields: to S of le Grand Fort, SSP.

Forts, listed by parish. See also individual entries.

SPP	Fort **George**, Fort **Irving**, Fort **de la Salerie**, Fort **de la Hougue à la Perre**, Fort **Souscription**
SSP	Fort **du Mont Crevelt**
Cdu V	Fort **Doyle**, Fort **Le Marchant**, **Star** Fort,
VdelE	none
C	Fort **des Grandes Rocques**, Fort **Hommet**
SSV	Fort **Richmond**
SPB	Fort **Saumarez** (l'Erée tower), Fort **Grey**
T	Fort **des Pezerils**
F	[Fort at le Gouffre, battery at la Corbière]
SM	Icart battery
SA	none

Fort à faire, le — land, fields: Havelet, SPP; les Roussiaux, C; Cravellet, SM; le Tertre, SA. Medieval earthwork to shelter animals. Misspelt **Fortes Affaires**, 18th c.

Fortebrie, de — house, land: Belval, CduV. Origin unknown. Recorded **Fortebrie**, 1654.

Fortescu, Fief de — feudal holding of land. SM

Fortin, de — field: Dobrée, SSP. After **Thomas Fortin**, 16th c.

Forval, du — fields, road: les Nicolles, F. Possible misspelling of **Torval** 'crooked valley'. q.v.

Fosse(s), des — houses, fields: Saumarez, Mare de Carteret, C; le Catioroc, le Haut, SSV; rue Rocheuse, la Francerie, les Sages, les Falles, le Bordage, SPB; les Jehans, les Tielles, T; la Villiaze, F; la Fosse, la Fallaize, les Rebouquets, SM. 'pit or ditch' for extracting minerals, or surname extinct 17th c. **De Fosse**.

Fosse, à l'**Arguille**, la — see **Arguille**.

Fosse Barrée, la — land: le Longtrac, SM. 'blocked ditch'.

Fosse, la **Platte** — field: la Vrangue, SPP. 'flattened pit'.

Fosses, (plural) with historical 'attached names', listed by parish. See also individual entries.

SPP	Fosse **André** (Landry)
SSP	none recorded
CduV	Fosse **Brehaut**
VdelE	none recorded
C	Fosse **Belengère**, Fosse **au Courly**, Fosse **des Guesdons** (**Gaidons**), Fosse **Job**, Fosse **à la Loe**, Fosse **à Sablon**.
SSV	Fosse **à l'Angle**, Fosse **d'Apbuisson**, Fosse **à l'Auge**, Fosse **au Chien**, Fosse **de l'Essart**.
SPB	Fosse **Cocquerel**, Fosse **de la Herbergerie**, Fosse **Tourteau Départy**
T	Fosse **au Larron**, Fosse **Piquenot**
F	Fosse **au Connillier**, Fosse **Denis Langlois**, Fosse **Robert Allez**
SM	Fosse **de Beaupeigne**, Fosse **de Bertrand**, Fosse **Mado**, Fosse **Paint**
SA	none recorded

Fossé, le — land: Chouet, CduV; la Crête, SPB; Pleinmont, T. 'bank' (hedgerow.)

Fossé, le **Tors** — land: le Crôlier, T. 'crooked bank'.

Fossé ès **Fayes**, le — land: Pleinmont, T. 'fairies' bank'.

Fosses, grandes — land: rue Rocheuse, SPB.

Fossette, la — fields: les Clospains, CduV; le Hamel, VdelE; le Bordage, les Sages, SPB. Diminutive of fosse (small pit).

Fouaillages, les — land: l'Islet, SSP; l'Ancresse, CduV; les Dicqs, VdelE. 'bracken patches' or 'scrub' on sandy waste land, for fuel.

Fouaschin, la **Terre** — land: la Couture, SPP.

Fouches, les	field: les Mielles, CduV. Probably extinct surname **Foche**. Recorded 1591.
Fouchin	fields: Ville au Roi, Mont Durant, la Couture, la Ramée, SPP; le Haut, SSV; la Vallée, le Campère, SPB; le Grée, la Planque, T. Surname **Fouaschin**, 17th c. **Fashion** 18th c. **Fachin** extinct 19th c. Also spelt **Fouachin**.
Fouchin, la **Rocque**	land: le Campère, SPB. Derivation as above.
Foullerez, le **Moulin**	site: S of le Foulon, SPP. former water mill (moulin à eau). Presumably connected with fulling cloth at le Foulon q.v . Abandoned and ruined 18th c.
Foullonnières, les	field: le Planel, T. patch of 'fuller's clay'.
Foulon, le	house, field: SPP/SA. 'fuller's clay'. Also a building, where cloth was fulled in the middle ages.
Foulon, douit du	stream: le Foulon, SPP.
Foullot, fontaine	spring: near Havelet, SPP.
Founin, le	field: les Petils, CduV. Surname extinct 15th c. **Sounil**. Recorded **Sounil** 1591, misspelt since 17th c.
Fouque, la **Hougue**	houses, fields: SSV. ' Fouque's hillock'. Surname extinct 17th c. **Fouque**. hougue q.v. Also feudal holding of land Fief des **Fouquées**, SSV.
Four, la **Frie** au	fields: SSV. Surname extinct 20th c. **Du Four**. frie q.v.
Four, le rue du	road: unlocated, VdelE.
Four Cabot, le	see **Cabot**.
Fourches, hougue des	land: N of Saumarez Park, C.
Fourches, le **Clos**	land: les Houmets, C.
Fourches, les	field: les Houmets, C.

Fournie, le camp	land: Mont Durand, SM.
Fourques, les	field: N of le Monnaie, SA. Records interchange the use of ch & q. 'forked'. Also spelt **Fourquet**. gibet q.v.
Fourques, rue des	road: unlocated, SA.
Fourrière, la	land: l'Hyvreuse, SPP; les Houmets, les Eturs, C; croûte au Cène, SPB; les Sages, T; la Fallaize, SM. Possibly a kiln.
Frairie, de	field: la Hougue Fouque, SSV. Probably owed rente to church fraternity. Also spelt **Frèrie**.
Franc Fief, le	see **Fief**.
Français, les	fields: N of la Hougue du Moulin,CduV; E of la Mare de Carteret, C; S of les Fauconnaires, SA. Surname extinct 17th c. **Le Francez**. Recorded **les Francois**, 1591, 1624.
Français, rue du	road: S of rue du Galaad, C.
France, de **Jean** de	fields: les Lohiers, les Frances, SSV. Derivation as above.
France, de **Pierre** de	fields: les Frances, SSV. Derivation as above.
France, de **Thomas** de	field: la Vallée, SPB. Derivation as above.
France, la **Croûte** de	field: les Monts (Delancey), SSP. Derivation as above.
Françerie, la	house, field: la Fosse, SPB. Derivation as above.
Frances, les	houses, fields: Candie, C; les Frances, le Courtil Ronchin, SSV; les Mouilpieds, SM; les Norgiots, SA. Surname extinct 19th c. **De France**.
Franche Terres, les	fields: l'Echelle, SA. Free of feudal dues owed on Fief Sainte Hélène.
Frazel, le **Clos**	field: les Houmets, C. Surname extinct 14th c. **Frasset**. clos q.v.

Frêne(s), le(s) — fields: les Monts (Delancey), SSP; les Eturs, C; les Fontenelles, F; la Croix Bertrand, la Fosse, SM; les Huriaux, SA. 'ash' (plantation).

Frênée, la — land: l'Hyvreuse, la Ville au Roi, SPP. Derivation as above.

Frênes, camp des — land: la Croix Bertrand, les Camps du Moulin, SM. Derivation as above.

Frênes, **Jardin** à — house: longue Rue, SSV. Derivation as above.

Frênes, rue ès — road: main road at la Fontaine, SSV. Derivation as above.

Frênes, ruette ès — lane: S-W of rue des Frênes, SM. Derivation as above.

Frères, la ruette des — land: alongside le Cimetière des Frères, SPP.

Frères, le **Cimetière** des — see **Cimetière**.

Frerie, de — see **Frairie**.

Frichet, le — see **Friquet**.

Frie(s), le(s) — houses, fields: les Fries, C; la Hougue Fouque, SSV; le Frie, SPB; le Frie, T; les Houards, F. Fallow allowed to revert to pasture for year-round grazing. Also spelt **Fricq**, **Fris**, **Frit**.

Frie au Four, le — see **Four**.

Frie Baton, le — see **Baton**.

Frie Plaidy, le — see **Plaidy**.

Frie Tourgis, le — see **Tourgis**.

Frie, le **Long** — house, fields: le Longfrie, SPB. Formerly a 'long' field. Derivation as above.

Frie, le **Mont** — land: Calais, SM. Derivation as above.

Fries, les **Plats** — field: les Vallees, C. 'flat' field. Derivation as above.

Friquet, le	houses, fields: la Bailloterie, CduV; le Friquet, les Grantez, C; la Hougue Fouque, la Longue Rue, SSV; les Maindonnaux, le Friquet, SM; le Camp Tréhard, les Bailleuls, les Naftiaux, SA. Originally unenclosed piece of fallow, uncultivated land by the roadside. Also spelt **frichet**.
Friquet au **Cure**, la	land: rue au Prêtre, C.
Friquet de la **Querruée**, le	land: near les Blancs Bois, VdelE.
Friquet des **Arguilliers**, le	land: le Grand Clos, C.
Friquet des **Buttes, le**	land: les Buttes, T.
Friquet des **Fauquets**, le	land: les Fauquets, C.
Friquet des **Grantez**, le	land: les Grantez, C. Site of fief court seat.
Friteaux, les	houses, land: N of la rue Poudreuse, SM. Formerly in fields. Plural of frit, a variation of frie q.v.
Froc(**q**), les	houses, fields: les Martins, SSP; rue Colin, VdelE; le Villocq, C. Surname extinct 17th c. **Du Frocq**, (**Du Fro**).
Frocq, la ruette du	lane: W of les Generottes, C. Derivation as above.
Frohiers, les	land: les Vallées, C. Surname extinct 14th c. **Frohier**. Misspelt **Forgiers**, 18th & 19th centuries. Also feudal holding of land, Fief **des Frohiers**, C.

G

Gabouret, de	field: les Mourants, SM. Surname extinct in Guernsey (extant Jersey 17th c.) **Gabourel**.

Gachères, les — house, fields: les Mouilpieds, SM. Recorded les **Gastières**, 1667. Possibly means 'fallow'.

Gagnaches, les — fields: les Martins, SPB. Surname extinct 15th c. **Gaignache**.

Gagnepain, de — field: N of les Grandes Capelles, SSP. Surname extinct 18th c. **Gagnepain**.

Gaidons, fontaine — spring: near Candie, C.

Gaidons, les — see **Dindons**. Also spelt **Guesdons**.

Gaillard, de — fields: rue Piette, C; Rue des Grons, SM. Surname extinct 15th c. **Du Gaillard**. Misspelt **Du Galliard**, 19th c. Also feudal holding of land, Fief de **Jean Du Gaillard**, SSV/SPB/T.

Gaillarde, la — fields: les Landes, C; la Villiaze, SA. Derivation as above.

Gain(s), le(s) — fields, road: W of St Briocq, SPB. Surname extinct 15th c. **Gain**.

Gal, le — fields: W of les Landes du Marche, VdelE; les Houards, F; Les Maindonnaux, SM; le Bouillon, SA. 'large boulder'.

Gal, camp du — see **Gal**.

Gal, fontaine — spring: W of les Landes du Marché, VdelE.

Gal, ruette du — lane: S of les Houards, F.

Gallet, le —

1. fields: les Gigands, SSP; les Annevilles, VdelE; les Houmets, C; rue Rocheuse, SPB; la Villette, SM. 'shingle' (pebbles).
2. foreshore: la Miellette, CduV; Mont-Chinchon, SSV; Braye-de-Lihou, la Rousse Mare (see **Les Anguillières**), SPB. 'shingle' (banks).

Gallet Heaume, le — foreshore: Havelet (now South Esplanade), SPP. Presumably '**Heaume's** foreshore'.

Galley, ruette du	lane: near la rue Rocheuse, SPB.
Gallez, au	fields: la Mare, le Videcocq, SPB. Extant surname **Le Gallez**.
Gallezeries, les	field: les Adams, SPB. Derivation as above.
Gallie, la	house, land, fields: les Rocquettes, SPP; rue des Marais, VdelE; la Rivière, C; les Hèches, SPB; E of les Laurens, T; Saints, SM; le Monnaie, SA. 'pebble patch', or extant surname **Gallie**.
Gallienne, Lorens	see **Laurens, Lorens**.
Galliennes, les	houses, fields: S of les Maingys, VdelE; N of les Adams, SPB; les Galliennes, T; les Merriennes, F; les Galliennes, SA. Extant surname, **Gallienne**.
Galliennes, rue des	road: unlocated, SPB.
Galliot, la **Rocque**	land: l'Ancresse, CduV. Derivation as above.
Galliot, le	fields: les Mielles, CduV; Icart, SM. Surname extinct 16th c. **Galliot**. Also feudal holding of land, Bordage **Galliot**, CduV.
Galliotte, la	land, fields: les Pelleys, C; Perelle, SSV; le Coudré, SPB; la Bouvée, la Quevillette, les Mouilpieds, la Galliotte, SM. Possible diminutive of **gallie**, 'fine pebbles'. Misspelt **Gaillotte**, **Gailliotte**.
Gand, de	site: medieval tower (unlocated) la Tourgand, SPP. 'place of safety'. tour q.v.
Garde, les **Maisons** de	former watchhouses: le Catioroc, SSV; la Prévôte (demolished), le Mont Hérault, SPB; Pleinmont (demolished) T; la Sommeillèuse (demolished) F; Icart SM. Built late 18th c. guet q.v.
Gardinet, le	house, fields: Côbo, le Gardinet, C; les Mourants, SA. 'small garden'. jardin q.v.

Garenne, la

1. land, fields: 'garende en la cloison du Valle', N of Vale Church, CduV. 'rabbit warren', circa 25 vergées. Enclosed warren on wasteland held by the abbots of Mont St. Michel.
2. land, moat: 'Garende D'Anneville', SSP; 'rabbit warren', about 8 vergées. Enclosed warren at Manoir D'Anneville built by Guillaume De Chesney, mid-13th c.

Garenne, **courtils** de la

land, fields: 'rabbit warrens'. Origin unknown, presumably privately owned.

1. land: N of **Landes du Marché**, SSP la **Garende**
2. fields: les Varendes, C les **Varendes**
3. field: Mont de Val, C la **Garenne**
4. fields: les Padins, SSV les **Guerandes**
5. fields: W of la Palloterie, SPB la **Varende**
6. fields: le Variouf, F la **Varende** (**Varende** from latin, warenna).

Garenne, la **Maison** de la

former house: Hougue Renouf, C. la Garenne 19th c. house name.

Gargatte, la

fields: les Bruliaux, SPB/F. 'gulley, gorge'.

Garis, de

house, fields: la Moye, CduV; Saumarez park, les Fauquets, C; les Tielles, T; rue au Cammu, SM; les Eperons, SA. Extant surname **De Garis**. Surname extinct 17th c. **Garie**. Also feudal holding of land, Vavassourie **Garrie**, CduV. hougue, messières q.v.

Garis, la **Hougue** des

land: les Eperons, SA. Derivation as above.

Garis, les **Messièrres**

land: la Haye du Puits, C. Derivation as above.

Gaspart, le

field: les Boulains, SSV. Forename (m) **Gaspard**.

Gaudin, le **Desert**

field: les Pages, SM. Surname (extant Jersey) **Gaudin**. desert q.v.

Gaudin, l'**Epine**	field: les Pages, SM. Derivation as above. epine q.v.
Gaudine, la	house: les Martins, SM. Derivation as above.
Gaulle, la	fields: St Jacques, SPP; la Croix Martin, le Gron, les Caches, SSV; les Brehauts, SPB; la Corbière, F. Possibly a variant of **gal**, q.v. 'boulder'.
Gautier, le **Mont**	cliff land: W of le Gouffre, F. Forename or surname, **Gaultier**. mont q.v.
Gauvain, de	fields: les Abreuveurs, SSP; le Hurel, SM; les Baissières, SA. Extant surname **Gauvain**.
Gauvainerie, la	house, land: les Merriennes, SM. Derivation as above.
Gazeaux, les	site: unlocated SM; 'franche prison', circa 1270 (Extente 1331).
Gazel, le	house, field: Saints, SM. Possibly extinct surname **Gassel**.
Géans du Roy	see **Roi**, les **Géans** du.
Geffrey, de	see **Giffré**.
Geffrey de Begueville, **Bouvée** de	feudal holding of land, dependency of Fief de **Blanchelande**, SM.
Geffrey, le **Bordage**	feudal holding of land, unlocated, SSP. Formerly belonged to **Ralph Geffrey**, 1309. Bordage q.v.
Gehou, de	see **Jethou**.
Geine, au	see **Jaignes**.
Gelé, le	houses, fields: le Marais, C; les Padins, SSV; la Houguette, la Pomare, SPB; les Niaux, Four Cabot, SA. Surname extinct 17th c. **Le Gelley**, (after **Michiel Le Gelley**, 1504, **Collas Le Gelley** 1610).
Geliennes, les	see **Juliennes**.

Genâts, à Vallée land: les Niaux, SA. Surname extinct 16th c. **Du Genat**.

Genâts, les houses, fields: les Genats, C; le Monnaie, SA. Derivation as above.

Genâts, rue des road: E of route de Carteret, C.

Genemye, de see **Jénémie**.

Generotte, la fields: Mont Plaisant, C; les Domaines, SSV. Diminutive of **genêt**, 'broom'.

Generotte, ruette de la road: W of la Haye du Puits, C.

Génêtier, le fields: le Dehus, CduV; le Pont, SPB; Pleinmont, T. Area of 'broom'. Also spelt **Généquet**, **Génétel**, **Génétière**, **Générstiere**.

Génêt(**s**), à land, fields, all parishes.

Génêts, la croûte a land: W of Queen's road, SPP.

Geniesse, la houses, fields: F. 'heifer', i.e. grazing for young livestock. Recorded **Ginesse**, 1671.

Gentez, le field: le Variouf, F. Origin unknown. Spelt **Jentez** until 1938.

George, Fort fort: citadel, SPP. After king George III.

George, la **Chapelle St** see **St George**, la **Chapelle**.

George, la hure land: le Mont Herault, SPB.

George road road: la Bertozerie, Les Godaines, SPP. Derivation as above.

George, **Saint** house, land: St George, C. Named after the 12th c. chapel, which was used as the court house of Fief le Comte after the Reformation.

George street street: Neuville (Newtown), SPP. Early 19th c. After the developer, Sir George Smith.

Georget(**s**), le(s) fields: la Gallie, les Menages, SPB. Forename(m) **Georget** (after **Georget Massy**, 1492).

Gérémie, de — field: la Moye, CduV. Misnaming of 'courtil **Sablon**', 1922. sablon q.v.

Germain, de — field: les ruettes Brayes, SPP. Surname extinct 17th c. **Germain**.

Gersiez, au — see **Bernabé**, **Ferenbras le Gersiez**.

Gervais, clos du — see **Gervaise**.

Gervaise(**s**). le(s) — fields: la Planque, C; les Jaonnets, SSV. Surname extinct 16th c. **Gervaise**. clos, q.v. Also feudal holding of land, Fief des **Gervaises**, SSV.

Get, au — see **Jet**.

Gethehou, de — fields: la Pitouillette, C; unlocated SSV. Surname extinct 17th c. **De Gethou** (after **Philipin De Gethou**, 1555).

Gibauderie, la — houses, land: N of les Rocquettes, SPP. Extinct surname, **Gibaut**.

Gibel, le **Mont** — mount or promontory: quarried for **Commercial arcade** SPP.

Gibet, le — field: les Fourques, SA. Medieval gibet (scaffold) for hanging malefactors.

Giffard, de — fields: les Huriaux, SM. After **Richard Giffard**, 1555. Extant surname **Giffard**. Also feudal holding of land, Bordage **Giffart**, CduV.

Giffard, rue — field: les Grantez, C.

Giffardière, la — fields: E of Albecq, C. Derivation as above.

Giffré, camps — see **Giffré**.

Giffré, le — fields: les Hures, CduV; les Hougues, SSP; les Rouvets, SSV. Forename (m) **Guiffrey**, also spelt **Geffrey**, **Guiffré**, **Gifrey**, **Jeffrey**.

Gigand de la Rue — former field: les Canus, SSP. Early 19th c. name. Recorded **Gingant de la Rue**, 1801.

Gigands, les — house, fields: les Canus, SSP. Probably extinct surname **Gingant** (les **Geingans**, 1663).

Gilloitte, la	see **Girouette**.
Giolle, la **Longue**	fields: les Rouvets, VdelE; La Landelle, C. Origin unknown. planque q.v.
Gion, le **Clos**	field: Sous l'Eglise, SSV. Extinct surname **Gion**. Also spelt **Guion**. clos q.v.
Giot, de	field: les Landes, C. Surname extinct 16th c. **Giot**.
Girard(s), les	house, fields: les Girards, les Vallées, C; Farras, F. After **Guillaume Girard**, 1554. Extant surname **Girard**.
Girard, de **Marie**	field: les Touillets, C. Derivation as above.
Giret, de	field: Bordeaux, CduV. Forename (m) **Giret**.
Girouette, la	house, land: les Rouvets, SSV. Recorded as above 1723 to present. Some records also give the name as **Gilloitte**, **Guilloitte**, 1718 to 1854. Possibly associated with a weathervane.
Givets, les	field: Cocagne, CduV. Origin unknown. Recorded **Guiais** 1654, **Gives** 1755.
Glagelier(**s**), le(s)	field: Hacse, CduV. Origin unknown.
Glagolière, la	field: les Barrats, VdelE. Origin unknown. Misspelt **Glazeliers**, 20th c.
Glategny, à	houses, land: SPP. Probably extinct surname **Glatigny**. croix, port q.v.
Glayeuls, les	land: la Croisée, F. 'wild iris'.
Glié, la	land, fields: les Terres, SPP; le Bigard, F. 'stubble'. mont q.v.
Gobtel, le	fields: les Monts, SSP; les Beaucettes, CduV. Surname extinct 16th c. **Le Gobetel**.
Godaines, les	houses, fields: Havelet, SPP. Surname extinct 14th c. **Godeyne**.
Godaines, rue des	road: W of George Road, SPP. Now called an avenue.

Goddards, les	houses, land: rue Etur, C. Surname **Goddard**. rue Etur q.v.
Gode, le **Camp**	see **Code**.
Godefroy, le(s)	fields: Port de Nermont, CduV; le Crocq, SPB. Forename (m) or surname **Godefroy**. Also spelt **Godfray**, **Godfreys**, **Godfroys**.
Godefroy, la **Croix**	site: les Landelles, C. Former wayside cross. Derivation as above.
Godefroy, la hure	land: W of la Hougue Anthan, SPB.
Godefroys, rue de	road: now les Clercqs, SPB. Derivation as above.
Godinette, hougue de la	land: S-E of les Hures road, CduV.
Godinette, la	field: unlocated. CduV. Origin unknown. hougue q.v.
Godios, les	fields: les Capelles, SSP. Plural of extinct surname **Godel**.
Gohier, camp de la **Rocque**	see Gohier, la **Rocque**.
Gohier, la **Rocque**	land: Houle au Lievre, (le Marais), C. Extinct surname 14th c. **Gohier**. rocque q.v. Also feudal holding of land, Fief des **Gohiers**, SSV.
Golle, la	see **Gaulle**.
Gonde, le **Val**	land: Jerbourg, SM. Origin unknown. Val q.v.
Gonebec(q), de	land: Candie, C. Extinct surname **Gonbec**.
Gonebec, fontaine	spring: near le Friquet, C.
Goosse, de	field: le Hamel, VdelE. Origin unknown, earliest record 1807.
Gosselin(s), le(s)	land, fields: Ville au Roi, SPP; le Douit, VdelE; les Emrais, C; le Variouf, F. Extant surname **Gosselin**. terre q.v. Also feudal holding of land, Fief **Robert Gosselin**, VdelE.

Gosselin, les **Terres**	land: Belvedere, SPP. Derivation as above.
Gosselinerie, la	house: former name of house, Grands Moulins, C. Reference in 1818. Derivation as above.
Gosselin, la croûte	land: Ville au Roi, SPP; la Hougue du Pommier, C.
Gosses, les	field: la Mare de Carteret, C. Surname extinct 17th c. **Gosse**.
Gots, la rue à	fields, lane: C. 'boulders'. Probably taken from le Déhuzet q.v. nearby. rue q.v.
Gouais, la rue au	road: la Ville Amphrey, SM. Former name of road. Surname extinct 17th c. **Gouie**, **Le Gouie**, also spelt **Le Gois**, **Le Gouais**.
Goubeys, les	house, fields: la Ramée, SPP; Grandes Mielles, VdelE; le Gron, SSV. Surnames, **Des Goubais** extinct 17th c., **Le Goubey** extinct 19th c. Also feudal holding of land, Fief du **Quartier des Goubeys**, VdelE.
Gouffre, fortification	earthwork: W of Petit Bot, F.
Gouffre, le	land: S of les Fontenelles, F. 'abyss, gorge'.
Gouie(s), le(s)	house, land, road: rue au Gouies, Ville Amphrey, SM; E of les Eperons, SA. Derivation as above. Also feudal holding of land, Fief **Gouie**, SSV.
Gouies, le **Marais**	fields: N of route de Plaisance (Airport), SSV. Derivation as above. marais q.v.
Gouvainerie, la	see **Gauvainerie**.
Gracien, de	fields: Bouet, SPP; Hougue des Haies, VdelE. Forename (m) **Gracien**, also spelt **Gratien**.
Grais, le	house, fields: W of les Reveaux, SPB. 'gravel, clay'. Le Grée q.v.
Grais, rue du	road: unlocated, SPB.
Grand(e)	'grande' label often attached to substantive name to indicate a big field of that name.

Grand(e) 'large, big' with proper names listed by parish (see also main text).

- SPP — Grand **Bosq**, Grand **Carrefour**, Grande **Croûte**, Grande **Marche**, Grand **Ménage**, Grande **Rue**.
- SSP — Grand **Fort**, Grandes **Jaonières**, Grande **Maison**, Grandes **Maisons**, Grand **Pont**.
- CduV — Grand **Camp**, Grand **Havre**, Grande **Lande**, Grand **Mains**, Grande **Maison**, Grande **Rue**.
- VdelE — (none of significance)
- C — Grande **Censière**, Grand **Courtil**, Grande **Croûte**, Grande **Hougue**, Grand **Lot**, Grand **Mare**, Grand **Marais**, Grands **Marchants**, Grands **Moulins**, Grand **Mourain**, Grand **Pré** (2), Grandes **Rocques**.
- SSV — Grands **Courtils**, Grand **Douit**, Grande **Lande**, Grand **Ménage** (3), Grands **Prés**, Grandes **Ravines**, Grande **Rue**, Grande **Pièce**.
- SPB — Grands **Camps**, Grandes **Fosses**, Grand **Menage**, Grandes **Rues**.
- T — Grande **Maison**, Grand **Pièce**. Grand **Hâtenez**.
- SM — Grand **Chemin**, Grande **Rue**, Grande **Pièce**.
- SA — Grand **Belle**, Grands **Courtils**.

Grand Camp, croûte du — land: near Chouet, CduV.

Grand Camp, houguette du — land: unlocated, CduV.

Grande Maison, la — see **Maison**.

Grandes Maisons, les — houses, land, road: to W of parish church, SSP. Late property of Le Marchant family, early 19th c. Derivation from family name **Grandmaison** (Grantmaison) q.v.

Grandin, la **Hougue** — field: le Dos d'Ane, C. Extinct surname (extant Jersey) **Grandin**. hougue q.v.

Grandineries, les — field: Fort George, SPP. Derivation as above.

Grandmaison, de — land: CduV. Surname extinct 16th c. **Grantmaison**. Also spelt **Grantmesson**,

	Grandemaison. Also feudal holding of land, Vavassourie **Grandmaison**, CduV.
Grand Mare	see **Fee Farms**.
Grand Moulin, douit du	stream: Talbot Valley, C.
Grand Port	mooring anchorage: Rocquaine Bay.
Grands Moulins, les	water mills: S of la Grande Mare, C. The district is known colloquially as King's Mills. 1. le Grand (Main) Moulin. At main road junction, formerly the King's Mill (on Fief St. Michel), the principal mill. Now a mains water pumping station. 2. le Moulin du Milieu. To the S, now a house. 3. le Moulin de Haut. Further to the S at a valley fork, intact and now a house.
Grandes Rocques, Fort	fort: N end of Cobo Bay, C.
Grange, la	1. houses, land: Grange road, SPP. Formerly 'grange' or barn of the Crown at **le Manoir de la Grange**. 2. field: la Vrangue, SPP 3. houses: route du Braye, la Hurel, CduV; la Porte, C; les Huriaux, le Douit, SSV; la Pomare, SPB. manoir q.v.
Grange De Beauvoir, la	house, fields: SPP. late of **Richard De Beauvoir**, 17th c.
Granges, les	house, land: la Porte, C. Presumably connected with le Manoir de Ste Anne q.v.
Granges, les **Camps** à la	fields: les Blanches, SM. Presumably near former barn of Fief de Sausmarez, SM.
Grangiers, Fief des	feudal holding of land. Dependency of Fief des Vingt Bouvées du Villain Fief le Comte. C.
Grante, de	fields: les Montmains, CduV; Bullaban, VdelE; les Landes, C. Surname extinct 18th c. **Grente**. Also spelt **Grantes**, **Grentez**.

Grantez, les	fields: les Grantez, C. Derivation as above.
Grantez, le **Moulin** des	site: rue de la Haye, C. wind mill. Demolished in 1940s during German Occupation. Derivation as above. Moulin à vent q.v. Also feudal holding of land, Fief des **Grantez**, C.
Grantez, rue des	road: N of les Jaonnets, C.
Grassin, de	field: le Bordel, CduV. Origin unknown. Recorded **Grassin**, 1755.
Grat, le **Port**	bay: SSP. Contracted form of Port **Vaugrat**. Port, Vaugrat q.v.
Gratien, houguette	land: le Retot, C.
Gratis, hougue	land: Sous les Courtils, C.
Gravée(s), la, les	house, former fields: rue de la Gravée, SPP. 'notched'. Origin unknown.
Grée, **le**	house, fields: N of route de Pleinmont, T. 'gravel clay'. Also spelt le Grais.
Gregoire, hougue	land: unlocated. CduV.
Grellichons, les	land: Fief le Comte (Felconte), SPB. 'lumpy'.
Grente,	see **Grantez**.
Grentez,	see **Grantez**.
Grés, le	house, land: le Gron, SSV.
Grés, le	house, land: E of le Courtil Ronchin, SA. 'gravel clay'. Also spelt Grais, q.v.
Grève, la	house, fields: N of la Ville Baudu, CduV. 'sandy beach'. Also sandy areas in Island bays.
Grève, la **Belle**	formerly a fine, sandy beach: les Banques, SPP/SSP.
Grey, **Fort**	fort: centre of Rocquaine Bay. SPB. Formerly Rocquaine Castle, rebuilt and renamed in 1804.
Griffon, hougue	see **Griffon**.

Griffon(s), le(s) — fields: les Arguilliers, VdelE; la Vieille Rue, SSV. Surname extinct 17th c. **Griffon**.

Gringoire, de — fields: Baubigny, SSP. Surname extinct 16th c. **Gringoire**.

Gringoire, le **Douvre Colin** — headland: Fief le Comte (Felconte), SPB. Derivation as above. Douvre q.v.

Gripé, camp du — land: rue des Renouards, C.

Gripel, le — land: Pleinmont, T; les Mouilpieds, SM. Surname extinct 15th c. **Gripel**.

Grippios, les — houses, fields: E of la Hougue du Moulin, CduV. Plural of **Gripel**. Derivation as above.

Griquettes, les — see **Criquettes**.

Griseaux, les — fields: rue Rocheuse, SPB. Plural of extinct surname **Grisel**.

Groignet, clos du — land: unlocated. C. Also spelt **Gronnet**.

Groignet, le — house, fields: le Groignet, le Marais, C. Probably unrecorded extinct surname **Gronnet**. Also feudal holding of land, Fief du **Groignet**, C.

Groin, **Bouvée** du — feudal holding of land. Dependency of Fief Gohiers, C.

Gron(s), le(s) — houses, fields: la Ramée, SPP; le Gron, SSV; E of la Croix Foret, F; la Villette, ruette Rabey, SM. Surname extinct 15th c. **Le Groing**.

Grons, fontaine des — spring: E of St Catherine at la Ramée, SPP.

Grons, la rue des — road: W of Forest church, formerly la **Voie de l'Egise**, now called le **Vue de l'Eglise**, F; E of la Villette, SM. Derivation as above. Also feudal holding of land, Bouvée du **Gron**, (Groing) C.

Gros, clos au — land: E of Saumarez Park, C.

Gros, fontaine — spring: near la Ramée, SPP.

Gros, le(s)	fields: la Ramée, SPP; rue Cohu, C; SA. Surname extinct 16th c. **Gros**, **Le Gros** (extant Jersey). Also feudal holding of land, Fief **Massy Gros**, SM.
Gros Mont, le	see **Mont**.
Grosse Hougue, la	see **Hougue**.
Grosse Pierre, la	see **Pierre**.
Grosse Rocque, la	see **Rocque**.
Gruchy, de	field: les Houmets, C. Surname **Gruchy** (**De Gruchy**) (extant Jersey).
Grunios, les	field: Hougue Pierre, CduV. Origin unknown. **Grinos**, 1591. Also spelt **Grunieaux**, 18th c.
Grut, de	fields: le Foulon, SPP; le Douit, VdelE; les Fauconnaires, SA. Extant surname **Grut**. Spelt **Gruth**, 1504 until late 17th c.
Grutte, la	field: la Gallie, SPB. Owned by Thomas **Grult**, 1549.
Guelle, la	former fields: rue des Guelles, SPP. Extinct surname **Guelle** (**De Guelle**, extant Jersey). Also spelt **Gueille**.
Guerande, la	see **Garenne**.
Guerbazel, le	land: Hougue Juas, CduV. Origin unknown. Recorded **Garbasel** 1591.
Guérin, de	field: le Tertre, SA. Extant surname **Guerin**.
Guérin, la **Croix**	site: wayside cross, junction of rue Maze and la Grande Rue, SM. Derivation as above.
Guerrier, le	see **Querrier**.
Guesdons, fosse	land: Candie, C. Also spelt Gaidons.
Guesdons, les	see **Dindons**.
Guet, le	see **Garde**, les **Maisons** de.
Guiffré, **Guiffrey**	see **Giffré**.

Guignon, de	houses, fields: rue Maze, SM; Fermains, Mont Durand, la Fallaize, SM. Surname extinct 18th c. **Guignon**.
Guignon, de **Martin**	field: la Bouvée, SM. 17th c. owner **Martin Guignon**. Derivation as above.
Guignon, la croûte	land: near Maurepas Road, SPP.
Guilbert, de	fields: le Mont Morin, SSP; la Grande Mare, C; les Buttes, SSV; les Marchez, SPB. Extant surname **Guilbert**. Spelt **Guillebert** prior to 17th c.
Guilbert, le **Courtil**	house: La Palloterie, SPB. Now called les Guilberts. Derivation as above.
Guillard, de **Pierre**	field: rue de la Cache, C. Extant surname **Guillard**.
Guillard, la **Terre**	land: les Baissières, SPP. terre q.v.
Guillaume, de	field: le Coudré, SPB. Forename (m) **Guillaume**. Called courtil **William** prior to 1807.
Guillaume Du Val, de	see **Val**.
Guillaume le Mière, fontaine	spring: le rue du Marais, C.
Guillaumes, les **Grand**	house, land: W of les Vauxbelets, SA. Origin unknown. Earliest record 1845.
Guille, de	fields: Fosse André, la Couture, SPP; le Marais, VdelE; le Camp, C; les Houards, F; les Eperons, les Baissières, SA. Extant surname **Guille**, derived from **Guillaume**, **Guillome**.
Guille, la **Croûte**	field: les Houards, F. Derivation as above.
Guillemet, la **Terre**	land: unlocated F. Forename (m) **Guillemet**. terre q.v.
Guillemette, de	fields: Airport, F. Forename (f) **Guillemette**. Also spelt **Guilmette**.
Guillemette, la **Croix**	site: N of les Hougues (le Dos d'Ane), C. wayside cross.

Guillemotte, de	fields: les Rohais, SA. Surname extinct 17th c. **Guillemotte**.
Guillemotte, la **Vallée**	land: l'Echelle, SA. Derivation as above.
Guilloitte, la	see **Girouette**.
Guillon, de	field: les Beaucamps, C. Forename (m) **Guillon**.
Guillon, la **Croix**	site: N of les Rebouquets, SM. wayside cross.
Guillote, de	field: rue d'Aval, VdelE. Forename (f) or surname **Guillot**.
Guillotin, de	field: Cocagne, CduV. Forename (m) **Guillotin**.
Guillot Justice, Fief de	feudal holding of land. T.
Guilmette, la	see **Guillemette**.
Guilmine, de	field: Cauchebrais, VdelE.
Guilmine, la **Hougue**	house, field: Juas, CduV. Forename (f) **Guillemine**.

H

Hache, la	fields: le Friquet, CduV; la Rivière, C; Cinq Verges, les Jaonnets, SSV; les Blicqs, F; Chemin le Roi, SM. Probably a form of héche q.v.
Hache, la **Plate**	field: les Etibots, SPP. Origin probably as above.
Hacsé, de	fields: CduV. Probably extinct surname **Haquesais**. Recorded **Haquesias** 1654; **Hacces**,1755. Misspelt **Axcé**, 20th c. rue Mahault q.v.
Haddy, de	garden: les Grandmaisons, SSP. Surname extinct 14th c. **Hady**.
Hagiers, les	house, field: Vaugrat, SSP. Surname extinct,

Le Hageys. Also spelt **Haguiers**, **Haguets**.

Haie, la **Hougue** — land: les Beaucettes, CduV.

Haie, sur la — fields, land: les Grantez, C; le Mont Durand, SM. 'hedge (bank)'. Also spelt **Haye**.

Haies, **Camp** des — land: Mt. Durand, SM.

Haies, la **Hougue** des — fields: la Charruée, VdelE. Surname extinct 16th c.**Hais**, extant **De La Haye**. Haisette, haie, hougue q.v. Also feudal holding of land, Fief du **Camp des Haies**, VdelE.

Hailla, de — former fields: Ruettes Brayes, SPP. Recorded **Fleurie Hailla**, 16th c. Surname extinct 13th-14th centuries, **Harla**. Misspelt **Hallia**, 19th c. fleurie q.v.

Hailla, la **Hougue** — fields: St Germain, C. Derivation as above. hougue q.v.

Hailla, la **Houguette** — land: rue des Toullez, C.

Hailla, la **Mare** — land: les Salines, SPB. former pond. Derivation as above. mare, saline, vivier q.v. Also feudal holding of land, Fief **Hailla**, SM.

Haillerie, la — fields: la Pomare, SPB. Derivation as above.

Haillia, de — see **Hailla**.

Haissettes, les — fields: rue Rocheuse, SPB. Diminutive of haize, low hedge bank or barrier.

Haize, la — fields: le Tertre, Cdu V; rue des Auberts, N of Le Bourg, now in Airport, F. 'hedge (bank)'.

Hallouvris, de — field: (unlocated), SA. Surname extinct 19th c. **Hallouvris**.

Hamel, le — house, fields: W of Pleinheume, site of **Chapelle de Notre Dame de l'Epine**, VdelE; E of Richmond, SSV. Extant surname **Hamel**. chapelle, epine q.v.

Hamelin, la **Croûte Renovet**	land: St Magloire, CduV. Surname extinct 19th c. **Hamelin**. croûte q.v.
Hamelin, la **Fontaine**	watering place: rue Poudreuse, SM. Derivation as above.
Hamelin, la **Rocque**	land: la rue Poudreuse, SM. Derivation as above.
Hamelin(**s**), le(s)	houses, fields: St Clair, SSP; les Hautes Landes, CduV; W of les Arquets, SPB; rue Poudreuse, SM. Derivation as above. rocque, fontaine, q.v.
Hamon, de	garden: le Chêne, F. Extant surname **Hamon**.
Hamonet, de	land: North side, CduV. Heirs **Hellier Hamon**, 1591. Derivation as above.
Han, au	fields: les Vicheries, le Mont Dril, SPB. 'galingale, sedge'. Also spelt **Hen**.
Hané, le **Pont**	see **Anne**, le **Pont**.
Hannière, la	fields, all parishes. Derivation as han above.
Hardouin, le	fields: la Croix Foret, F. Extinct surname **Hardouin**.
Hasios, Fief des	feudal holding of land. Dependency of Fief de Carteret, C.
Hâtenez, de	houses, fields: E of les Pièces, F. 'high or big nose', a feature on the landscape.
Hâtenez, le **Grand**	land: E of les Pièces, F. 'large' field. Derivation as above.
Hativet, le	house, fields: Baubigny, les Sauvagées, SSP. 'early cropping'.
Haule, Fief de la	feudal holding of land. Dependency of Fief le Roi, SA.
Haut, de	fields, houses, all parishes. 'top', relative to surroundings.
Haut des **Rouvets**	houses: les Rouvets, SSV. 'upper' houses, les Rouvets.

Haut, le — house: les Domaines, SSV. 'top' house in district.

Haut, **Hauts**, **Haute**, **Hautes** with 'proper' names are indexed under the substantive name;

Haute **Banque**, Hautes **Capelles**, Haut **Boullet**, Haut **Chemin**, Hauts **Courtils**, Hautes **Landes**, Hautes **Marais**, Haut **Pavé**.

Hauteur, la — houses: N of la Moye, CduV; W of la Hougue Anthan, SPB. 19th c. house name, 'height'.

Hauteville, à — district: outside S limit of medieval town, i.e. S of rue des Cornets and site of la **Tour Beaurégard**. 'upper town'. Recorded 1574.

Hautgard, le — site: St George, C. Former seigneurial rickyard of Fief le Comte.

Hautgard(**s**), les — house, fields: N of la Croûte Becrel, CduV. 19th c house name, the 'rickyards'.

Hautgard, le **Vieux** — field: Seaview farm, C. 'old rickyard'. A rickyard was an integral part of all properties. Also spelt **Haugart**, **Hogard**.

Haut Séjour — see **Séjour**.

Havelet, à — houses, land, district: E of Hauteville, SPP. Surname extinct 14th c **Havelet**. Also spelt **Havellet**. fallaize, hougue q.v.

Havelet, hougue de — see Havelet.

Havelet, **Ruette** de — land: near les Terres, SPP.

Havilland, (**vale**) de — district, houses, vale: N of les Damouettes, SPP. Extant surname **De Havilland**.

Havilland, de — fields: la Ramée, SPP; unlocated, VdelE; les Jaonnets, SSV; in Airport, F; les Hubits, SM. Derivation as above.

Havilland, **Douit** de — stream: Havilland Hall, SPP.

Havilland Hall — house: 19th c. mansion, le Vauquiédor, SPP. Built by Lt.Col. T. F. De Havilland on his return from India.

Havilland, la **Croûte** — see **Havilland Vale**.

Havilland, le **Manoir** de — site: les Vardes, (unlocated) SPP. Medieval manor recorded 1574. manoir q.v.

Havilland, rue — street: la Neuveville (New Town) SPP. Early 19th c. after Sir Pierre de Havilland, the developer

Havre de **Bon Répos** — see **Bon Répos**.

Havre, le — former field: S of harbour, SSP. 'haven'.

Havre, le **Grand** — large tidal haven: W of le **Braye du Valle**.

Haxcé, de — see **Hacsé**.

Haye, la rue — field: la Carrière, F. Presumably from extinct surname **Hais**. Haies q.v,

Haye du Puits, la — house: property S of Saumarez park, C. House recorded early 16th c. Earliest reference to name; **clos et haye du Puits**, 1663.

Heaume, au — field: les Houmets, C. Extant surname **Heaume**.

Heaume, de **Plein** — see **Pleinheaume**.

Heaumerie, la — house, land: Saints, SM. After **Pierre Heaume**, 1666.

Hèbergeries, les — see **Herbergeries**.

Héche(**s**), la(les) — houses, fields, 'gate' to arable, pastoral areas. Unnamed, approximate locations.

SPP	Grande Marche, la Pitronnerie, L'Hyvreuse, les Guelles.
SSP	St Clair, Baubigny, les Effards.
CduV	Communes de l'Ancresse (la Hache), Belval.
VdelE	rue d'Aval.
C	le Rocher, le Préel, les Grantez, les Beaucamps, les Houmets, les Hougues.
SSV	le Bélial, les Piques, la Vieille Rue, le Neuf Chemin, la Hougue Fouque.
SPB	la Devise, la Gallie, la Madeleine.
T	les Tielles.
F	les Landes.

SM	les Maindonnaux.
SA	les Rocquettes, les Maumarquis, les Fauconnaires, la Croix au Baillif.
Hechet, le	1. house, fields: S of les Hougues, C; la Pomare, SPB. 'lesser' gate. 2. windmill: Steam Mill lane, SM. 19th c. Also known as **Ozanne** mill.
Hemeril, le	fields: Hatenez, F. Surname **Hemery**. Emrais q.v.
Henris, la **Vallée** des	land: la Vieille Rue, SSV. Extant surname **Henry**. Also spelt **Hanry**, **Henri**, **Hanris**.
Henry, camp du	see **Henry**.
Henry, de	fields: les Pins, C; les Tielles, T; la Corbière, F. Derivation as above
Henry, de **Collas**	former field: la Vrangue, SPP.
Henry, de **Jean**	field: la Cache, C.
Henry, de **Pierre**	fields: les Grands Moulins, la Lande, C.
Henry, de **Thomas**	fields: les Adams, les Brehauts, les Sarreries, les Ménages, SPB.
Henry du Castel, Fief de	feudal holding of land. Dependency of Fief de Blanchelande, SM.
Héraults, les **Monts**	watchhouse, fields, promontory: SPB. Surname extinct 15th c. **Herault**. Garde, les Maisons de, mont q.v.
Herbager, le	fields: les Godaines, SPP; rue St Pierre, SSV; les Corbinez, SPB; les Landes Hue, la Hure, T. 'pasture'.
Herbe, à l'	fields: les Effards, Haye du Puits, C; Bas Courtils, SSV. 'grass'.
Herbe, clos a l'	see **Herbe**.
Herbergerie(s), la(les)	fields, fosse: les Cambrées, SPB. 'grassland'.
Herchiers, les	former fields: la Mare de Carteret, C. Presumably extinct surname, **Heriche**. Locally recorded, les **Harichis**, 1551.

Herclin Roche	see **Rocque, Roquelin**.
Herivel, la **Croûte**	former field: les Rohais, SPP. Extant surname **Herivel**. croûte q.v.
Hermitage, de	house: les Croûtes, SPP; les Maindonnaux, SM. 19th c. house name.
Hermon, le **Mont**	area: E of Queen's road, SPP. Early 19th c. name.
Herne, Fief	feudal holding of land. C.
Hérode, la **Ville**	see **Ville**.
Heron, de	fields: Fort road, SPP; les Maindonnaux, SM. After Collas Heron. Surname extinct 16th c. **Heron**.
Heron, la **Vallée** des	field: la Vieille rue, SSV. 'Mr Heron's valley'.
Heronnière, la	houses, land: les Banques, SSP; les Rébouquets, F. Derivation as above.
Hervy, de	fields: Bon Port, les Varinvarays, Jerbourg, SM. Surname extinct 16th c. **Hervy**.
Hervy, les **Messières**	land: Jerbourg, SM. messières q.v. Derivation as above.
Hervy, la **Lague**	fields, coastal area: la Lague, T. lague q.v. Derivation as above.
Heulle, le camp	land: rocque Gohier, C.
Heureux, le **Val**	valley: E of les Monts Herault, SPB. 'blessed', i.e.productive.
High Street	see **Rue**, la **Grande**.
Hillaire, Fief d'	feudal holding of land. Also known as Illaire. Dependency of Fief de Longues. SSV.
Hirzel, rue	street: SPP. Named after **Hirzel Le Marchant**, H.M. Procureur 1774 to 1792.
Hivreuse, l'	see **Hyvreuse**.
Hobodiaux, les	field: les Nouettes, F. Origin unknown.
Hocart, la **Pointe Henry**	land: les Bourgs, SA. Extant surname **Hocart**.

Hoguet, le **Clos** — house, fields: near les Prevosts, SSV. Surname extinct 16th c. **Hauguet**. Houiquet q.v.

Hommel, **Hommet**, le —
1. off-shore tidal islets.
   Castrum de Hommet now Château Cornet (Castle Cornet), SPP. **Hommet Benest** islet seaward of Bordeaux, CduV.
   Hommet Paradis islet seaward of le Port de Nermont, CduV.
2. land locked islets.
   Hougue du Hommet N of la Vrangue, SPP.
   Grand Hommet rabbit warren: W of le Château des Marais, SPP.
   Fort Hommet fortification: Martello tower, battery, former barracks, N of Vazon, C.
   Hommet de l'Eperquerie, le, land: Hougue de Nermont, CduV. Originally islet on foreshore. eperquerie q.v.
   Hommet Quenot, le, former field: rue des Marais, VdelE. Originally an islet in a marsh. Quenot q.v.
3. see **Houmets**, les, for place names derived from the surname **Du Hommet**.

Hôpital de la **Campagne**, l' — buildings, land: les Girards, C. 'Country Hospital', established 1752, now called Castel Hospital.

Hôpital de la **Ville**, l' — buildings, yards: le Truchot, SPP. 'Town Hospital', established 1743, now Police Headquarters.

Horaine, la **Croûte** — field: la Villiaze, SA. Presumably extinct surname. croûte q.v.

Hospital lane — street: adjacent to former Town Hospital, SPP. Formerly rue des Prestres. Prestre, rue q.v.

Houards, les — district: S of Forest church, F. After **John Ouar**, 1434. Surname extinct 15th c. **Ouar**, (**Houard**).

Houet(**s**), les — house, fields: Pleinmont, T; les Villets, F; les Boutefèvres, SM. Surname extinct 17th c. **Hue** (**Huet**, accentuated).

Hougue(**s**), la (les) — 1. 'unnamed' houses, land, fields, roads. 'hillock' (lookout) with approximate locations.

SPP	la Bertozerie, rue Poidevin, Glategny.
SSP	les Trachieries, les Hougues, Baubigny, les Gigands, les Nicolles.
CduV	le Tertre, Juas, Bordeaux, Paradis, les Hautgards.
VdelE	le Hamel, les Fortaitures.
C	Moulin de Haut, la Hougue, les Pins, le Douit, le Hechet, les Hougues.
SSV	l'Acbuisson.
SPB	le Coudré, le Catillon, la Claire Mare, St Briocq, la Gallie, les Sablons, les Paas, Rocquaine.
T	les Portelettes, les Laurens.
F	Mont Marché (in Airport).
SM	les Gachières.
SA	les Baissières, les Eperons.

Hougue à la **Perre**, **Fort** — fortification: S end of Belle Greve Bay, SPP.

Hougue, au **Son** — land: les Canichers, SPP. 'sounding'. Rocque qui sonne q.v.

Hougue Collin Denis, **Camp** de la — land: les Hommets, C.

Hougue de **Canicher**, la — land: les Canichers, SPP.

Hougue ès **Dames**, camp de la — land: les Pins, C.

Hougue Falle, **Ruette** de la — lane: la Hougue Falle, SPB.

Hougue, la **Grande** — land, fields: Longrée, CduV; les Pièces, F; les Mouilpieds, SM; les Blicqs, SA. 'great'.

Hougue, la **Grosse** — houses, fields: Saltpans, SSP. 'large, big'.

Hougue, la **Longue** — (reservoir): Bulwer avenue, SSP; Hougue Fouque, SSV. 'long'.

Hougue Ronde, la — land: le Hurel, CduV. 'round'.

Hougues, les **Petites** — houses: Hougue du Moulin, CduV. 'small'.

Hougues, les **Plates** — field: les Hauts Courtils, le Hamel, VdelE. 'flat'.

Hougues, rue des — road: S of le Dos d'Ane, C.

Hougues, **Sous** les — houses: la Hougue, Juas, Cdu V. 'below'.

Hougue(**s**), les — 2. with proper names listed by parish. See also individual entries.

SPP — Hougue **Beauvoir**, Hougue **de Canicher**, Hougue **de Havelet**, Hougue **Mourin**, Hougue **à la Perre**, Hougue **au Son**, Hougue **Vauvert**.

SSP — Hougues **Balan**, Hougues **Behier**, Hougue **de Bois**, Hougue **Lorenche**, Hougues **Magues**, Hougue **au Mière**, Hougue **des Martins**, Hougue **du Moustier**, Hougue **Nez** (**Noye**), Hougue **des Nicolles**, Hougue **de Noirmont** (de Nermont), Hougue **des Osmonds** (des Aumonts), Hougue **de Pulias**, Hougue **des Quartiers**, Hougue **des Romains**, Hougue **Samsonnet**, Hougue **Testart**, Hougue **des Trachys**, Hougue **Vaudin**.

CduV — Hougue **Bailleul**, Hougue **Bégine**, Hougue **Besnard**, Hougue **de Bordeaux**, Hougue **des Corvées**, Hougue **de la Godinette**, Hougue **Gregoire**, Hougue **Guilmine**, Hougue **des Hures**, Hougue **Jamblin**, Hougue **Jehannet**, Hougue **Juas**, Hougue **au Maitre**, Hougue **des Mares Pellées**, Hougue **du Moulin**, Hougue **Patris**, Hougues **Perres**, Hougue **Pinant**, Hougue **au Prêtre**, Hougue **Ricart**, Hougue **Robin**, Hougue **au Sage**, Hougue **Samson**, Hougue **à Suchez**, Hougue **du Valle**.

VdelE — Hougue **Béhehe**, Hougue **des Doreys**, Hougue **Griffon**, Hougue **des Haies**, Hougue **Rots**.

C	Hougue **des Cherfs**, Hougue **au Comte**, Hougue **de la Croix Guillemette**, Hougue **ès Dames**, Hougue **Collin Denis**, Hougue **Eri**, Hougue **des Fourches**, Hougue **Grandin**, Hougue **Gratis**, Hougue **Hailla**, Hougue **du Pommier**, Hougue **Renouf**, Hougue **Sotuas**.
SSV	Hougue **Bacheley**, Hougue **Fouque**, Hougue **Maban**.
SPB	Hougue **Anthan**, Hougue **Falle**, Hougue **d'Abraham Gallienne**.
T	Hougue **Taive**.
F	Hougue **Beausouffle**.
SM	[none]
SA	Hougue **au Boeuf**, Hougue **des Cambrettes**, Hougue **de Garis**, Hougue **au Messurier**, Hougue **du Moulin Foulerée**, Hougue **du Roi**, Hougue **Tourgis**.
Houguesin, le	fields: les Hougues, C. 'small knoll', part of a larger promontory.
Houguet, le **Vieil**	see **Hoguet**.
Houguette)s), la(les)	1. houses, fields, land. Diminutive of hougue q.v 'small' hillock. Listed by parish with approximate locations.
SPP	Rocque à l'Or, les Terres, les Granges.
SSP	la Saline, les Longs Camps.
CduV	la Rochelle.
VdelE	les Hauts Courtils.
C	les Houmets, la Robine, le Feugrel, le Carrefour, les Touillets, la Houguette.
SSV	la Grande Rue, les Jénémies, la Croix.
SPB	le Felcomte, la Houguette, les Sarreries.
T	[none]
F	l'Epinel, les Houards.
SM	la Grande Rue, la Fosse.
SA	[none]
Houguette(s), **les**	2. with proper names, listed by parish. See also individual entries.
SPP	Houguette **des Longs Camps**, Houguette **Collin**.

CduV	Houguette **Ballan**, Houguette **du Grand Camp**, Houguette **du Tor Camp**.
C	Houguette **Gratien**, Houguette **Hailla**, Houguette **au Tillier**.
SM	Chapelle **de St Jean de la Houguette**.
Houguette au **Tellier**, **Camp** de la	land: see **Tellier**, la **Houguette** au.
Houguette, **Camp** de la	land: les Granges, les Terres, Rocque a l'Or, SPP.
Houguettes, **Ruette** des	land: la Lande, C.
Houguillon, camp au	land: Camps Croix, SM.
Houguillon, le	fields: Croûte Becrel, CduV; la Valangot, SPB; le Rocher, T; les Nicolles, F; les Marettes, SM. Diminutive of houguette, 'very small knoll'.
Houiquet, le	see **Hoguet**.
Houle au **Catioroc**, la	field: vicinity of le Catioroc, SSV. 'hollow, crevice'.
Houle au **Lièvre**, la	house, land: le Hamel, SSV. 'Le Lièvre's hollow'.
Houlets, les	field: les Effards, C. Origin unknown. Spelt **Houles**, 1894. houle q.v.
Houlles, la **Vallée** des	land: les Effards, C. Houle, vallée q.v.
Houlot, le	fields: les Beaucamps, Clos Vivier, C. Diminutive of **Houle**.
Houmets, la rue des	road: les Houmets, C. Surname extinct early 16th c. **Du Houmet**. Also spelt **Du Hommel**.
Houmets, les	houses, fields: large area N of Saumarez park, C. Derivation as above. Current spelling for 'islets'. Hommel, hommet q.v.
Houmtel, le	house, field: N of Hougue du Moulin, CduV. Contracted spelling of **Hommete**l. Diminutive of **hommel** or **hommet**. 'tiny' islet.

Huberderies, les	fields: E of Pont Vaillant, SSP. 'late of **George Hubert**', early 17th c. Extant surname **Hubert**.
Hubert, la **Charrière**	field: la Bailloterie, CduV. 'track'. Derivation as above.
Hubert, le **Courtil**	field: les Adams, SPB. Derivation as above.
Hubert, le **Val**	land: les Tielles, T. Derivation as above.
Hubert, les **Vaux**	land: la Fallaize, SM. Derivation as above.
Hubits, les	district: N of les Blanches Pierres, SM. Surname extinct 16th c. **Le Huby** (**Huby** extant Jersey).
Huchettes, les	fields : les Mourains, Mont Plaisant, C; la Pomare, SPB. Probably extinct surname.
Huchettes, rue des	road: unlocated, C.
Huchon, le Clos	field: Richmond, SSV. Surname extinct 15th c. **Huchon**. Also feudal holding of land, Fief **Huchon**, SSV.
Hue, de	fields, land: Quay, SSP; le Bordel, CduV; Richmond, SSV. Forename or surname extinct 15th c. **Hue**. Houets q.v. Also feudal holding of land, Fief du **Domaine Dom Hue**, SSV.
Hue, camp du	see **Hue**.
Hue, la **Planque**	land: la Maraitaine, CduV. planque q.v.
Hue, le **Castel Jean**	land: les Monts Hérault, SPB. former 'earthwork'.
Hue, le **Moulin à vent** du **Domaine Dom**	see **Dan Hue**, le **Moulin**.
Hue, les **Vos**	field: E of le Courtil au Préel, SA. 'Hue's valley'. Also spelt **Voshues**, **Vohut**. val q.v.
Huelin, de	field: la Hougue de Haies, VdelE. Extant surname **Huelin**.
Huet, le **Courtil**	field: le Douit, CduV. First ref. 1837. Definition as above.

Huet, le **Moulin**	see **Luet**.
Huet, les **Champs**	fields: la Carriere, F. Extinct surname **Hue** or **Huet** (accentuated form).
Hugues, la **Terre**	land: la Couture, SPP. Surname extinct (in Guernsey) 16th c. **Hughes** (immigrant). terre q.v.
Huit Bouvées, Fief des	feudal holding of land. T.
Hung(u)ers, les	house, fields: les Lohiers, SSV; N of les Blicqs, SA. Surname extinct 15th c. **Hunger**. Also spelt **Hungiers**, **Hunguiers**.
Hunguier, clos	land: les Hunguets, SA.
Huray, au	field: la Mare, SPB. Extant surname **Le Huray**.
Hure(s), la(les)	houses, land, fields, roads. 'ridge'. Approximate locations listed by parish.
SSP	St Clair.
CduV	la Bailloterie, North side, Mont Crevel.
C	Candie, rue au Prêtre, les Beaucamps.
SSV	Sous l'Eglise, les Huriaux.
SPB	la Cache, la Prévôté.
T	la Hure, les Hures (both W of la Bellée).
SM	le Douvre, Bon Port, le Vallon.
SA	les Hunguets, le Grand Belle.
Hure, la **ronde**	field: la Croisée, T. 'rounded' ridge.

Hure(s) with proper names listed by parish. See also individual entries.

SPP	Hure **de la Hougue de Canicher**.
CduV	Hure **Mare**.
C	Hure **Corneille**.
SSV	Hure **Béco**.
SPB	Hure **Beuval**, Hure **George**, Hure **Godefrey**, Hure **de Lif**, Hure **au Mouton**, Hure **de St Briocq**.
T	Hure **des Camps Rulées**, Hure **du Crôlier**, Hure **Michel Dry**, Hure **de Letat** (de Letac), Hure **ès Jouppes**, Hure **de la Mare**, Hure **Valmette**.
SM	Hure **Baugy**.

Hurel, le	houses, fields, land. Diminutive of hure, shortened, minor ridge'. Approximate locations listed by parish.
CduV	Ville Baudu, le Hurel (la Grève), le Closel.
C	le Hechet.
SSV	les Ruettes, les Huriaux, le Hurel.
SPB	les Reveaux, la Rocque.
T	le Colombier, le Hurel (le Gain), la Forge, les Jardins.
SM	Sausmarez, Saints, le Vallon, le Mont Durand, Icart, le Hurel (le Vallon).
SA	le Hurel (le Bouillon).
Hurel de **Cravallet**, le	see **Cravellet**.
Hurel de la **Bette**, le	see **Bette**.
Hurel des **Brouards**, le	see **Brouards**.
Hurel, Du	Surname extinct 16th c. **Du Hurel**. Parish of St Martin. Possible connection with some place names in SM.
Hurel, fontaine	spring: le Vallon, SM.
Hurel, rue du	road: unlocated, V; W of le Vallon, SM.
Hures, **Hougue** des	land: Mont Cuet, CduV.
Hures, rue des	road: unlocated, T.
Hurette, la	fields, land: la Rochelle, CduV; les Rouvets, VdelE; les Niaux, la Hurette, C; la Ruette, les Nicolles, SM. Diminutive of **hure**, 'minor ridge'.
Hurette, rue de la	road: S of King's Mills, C.
Huriaux, les	houses, fields, land: les Huriaux, SSV; la Rocque, le Vallet, SPB; les Fontaines, T; Icart, les Huriaux, SM; les Rohais, SA. Plural of **hurel**. 'minor ridge'.
Huriaux, les **Camps** des	land: le Douit, la Rousse Mare, SPB.
Huyshe Memorial	building: opposite les Buttes, SA. Former Temperance café built in his memory by the widow of Maj. Genl. Huyshe, 1880s.

Hys, le **Clos**	field: Seaview farm, C. Surname extinct 16th c. **Thehy**. clos q.v.
Hyvreuse, l’	district: between Beau Sejour and the Fire Station, SPP. Origin of name unknown.
Hyvreuse, la **Pierre** de l’	site: the ‘**masse**’ of above windmill replaced the earlier standing stone (menhir). SPP. pierre, rocque q.v.
Hyvreuse, le **Moulin** de l’	site: ‘**masse**’ (mound) of former windmill which has Victoria Tower built on it (1846), SPP. Possible that the windmill was built in the mid-16th c.

I

Icart, d’	land: promontory, Icart, SM. fields: la Villette, les Mourants, SM; les Bémonts, SA. Surname extinct 16th c. Heirs **John Dicart**, 1574. Also **Discart** (see Dixcart, Sark).
Icart, **Camp** en	land: Icart, SM.
Icart, **Fort**	fort: between Saints Bay and Petit Bot Bay, SM.
Illaire, Fief d’	see **Hillaire**, Fief d’.
Image, l’	fields: les Pièces, SM. Origin unknown. Recorded 1671.
Imbert, la **Banque**	foreshore: Bordeaux, CduV. Probably extinct surname **Imbert**.
Ingot, le **Val**	fields, land: S of les Portelettes (on the crest), T. Surname extinct 15th c. **Ingot**. Also spelt le **Valingot**.
Ingot, la **Vallette**	land: le Gain, SPB. Derivation as above.
Irving, **Fort**	fort: part of Fort George, les Terres, SPP. After Major General M A Irving, Lieut-Governor in 1780.

Isabelle, la **route**	road: les Croûtes, SPP. 20th c. construction
Islets, les	houses, fields: N-W of le Longfrie, SPB. Surname extinct 16th c. **Des Isles**. Confused in 17th c with **De Lisle**, later owners of the property.
Islets, **Moulin** des	site: E of Arsenal drill hall (States houses), SPB. Formerly a windmill, moulin à vent. Derivation as above. Arsenal, Maison des Pauvres, moulin à vent q.v.
Isory, d'	see **Orais**.
Israel Le Lacheur, le **Douit**	stream, field: Douit d'Israel, SPB; la Forge, T. After **Israel Le Lacheur**, owner 1626. douit q.v.
Issue(**s**), l'(les)	houses, fields, land: La Hurette, C; les Issues, la Hougue Fouque, SSV; les Issues, SPB; les Cambrées, T. strip of land for 'exit'.
Ivelin, la **Croix**	see **Yvelin**.
Ivy castle	castle: 18th c. name for Château des Marais. Chateau, marais q.v.

J

Jacques, de	fields: les Landes du Marché, Pleinheaume, VdelE; les Caches, SM; in Airport, SA. Forename (m) **Jacques**.
Jacques, la **Chapelle St**	see **St Jacques**, la **Chapelle**.
Jacques, la **Pièce St**	see **St Jacques**, la **Pièce**.
Jacques, **St**.	see **Saint Jacques**.
Jaigne(**s**), au(des)	fields: les Valettes, la Hougue Fouque, SSV; la Villiaze, in Airport, SA. Presumably extinct surname **Le Geyne**. Also spelt **Gene**, **Geine**, **Jaines**.
Jaignes, la rue des	road: les Vallettes, SSV. Derivation as above.

Jamblin, de land: le Closel, CduV. Extinct surname **Gaubelin**. Recorded **Gaubelin 1591**. Also spelt **Jambelin**. hougue q.v.

Jamblin, **Hougue** land: S of Jamblin road, CduV.

James, de field: la Vrangue, F. Forename (m) **James**.

Jamette, de fields: les Etibots, VdelE; les Vinaires, SPB. Forename (f) contraction of **Je**(**h**)**annette**, also spelt **Jennette**.

Jamin, le **Clos** field: les Padins, SSV. Extinct surname **Jamin**.

Janin Besnard, Fief feudal holding of land. SPB and T.

Jannel, **Janneaux**, le(s) see **Jaonnet**.

Jaon(**s**), a(des) fields, land: le Catioroc, SSV; la Pointe, SPB; le Planel, T; le Varclin, Saints, SM. 'furze, gorse'. Also spelt **Janc**(**s**).

Jaon, **Camp** à Le Varclin, SM.

Jaonnet(**s**), le(s) houses, fields: (le Jannes) Retôt, la Masse, les Jaonnets, C; les Jaonnets, SSV; (le Jannet, les Janneaux) le Jaonnet, Saints, SM. 'small' areas of furze, gorse. Also spelt **Jannes**, **Janneaux**, **Jeannes**.

Jaonnets, **Camp** des land: les Jaonnets, C.

Jaonnets, rue du road: E of route des Houguets, C. Now called rue des Boulains.

Jaonnette, la fields: (unlocated) CduV; Pleinmont, T. Alternative to Jaonnet q.v.

Jaonneuse, la land: Chouet, CduV. Recorded **Janneuse** 1775, a variation of **Jaonnet**.

Jaonnière, la 'furze brake', all parishes. An enclosed area of furze, gorse, kept for cutting for fuel every third year.

Jaonières, **Grandes** land: Bulwer Avenue, SSP

Jaons du Roi, les see **Roi**, les **Géans** du.

Jaquet, de — field: les Houards, F. Forename (m) **Jacquet**.

Jardin(s), le(s) — houses, fields: Les Genâts, les Pelleys, C; la Lague, SPB; les Brouards, T. 'the garden(s).' Gardinet q.v.

Jardin à Frène, le — see **Frène**.

Jardinets, les — see **Gardinet(s)**.

Jardin Renault, le — see **Renault**.

Jardins Blicq, les — see **Blicq**.

Jean du Gaillard, Fief du — feudal holding. SPB and T.

Jean Fallaize, **Fontaine** — spring: Mont Durand, SM.

Jean Toulle, la **Croix** — cross: les Touillez, C.

Jeannes, les — see **Jaonnets**.

Je(h)an, de — as a forename and used with a surname is indexed under the substantive part of that surname. Jean **Allaire**, Jean **de Beauvoir**, Jean **Brehaut**, Jean Le **Gouais**, Jean **Henry**, **Catel** Jean **Hue**, Jean **Le Lacheur**, Jean **De Lisle**, Jean **De La Marche**, Jean **Mollet**, Jean **Moullin**, Jean **Des Perques** (**Desperques**), Jean **Toulle** (**Touillet**).

Jehan, de — as a surname, see below.

Jehan, de **Jean** — field: rue Rocheuse, SPB. Sale by son **Collas Jehan** recorded, 1532.

Jehan, de **Marie** — field: les Buttes, SSV.

Jehan Denis, **Fontaine** — spring: unlocated. SA.

Je(h)anne Collin, la **Croix** — see **Collin**, **Croix**.

Jehannet, de — fields: Cognon, CduV; Saumarez, C; la Villette, rue Poudreuse, SM. Forename, **Jehannet** ('little John'). Also spelt **Joinet**, q.v.

Jehannet Baudain, de — see **Baudain**.

Jehannet, la **Hougue**	former field: E of la Hougue du Valle, CduV. Derivation as above. hougue q.v.
Jehannette, de	field: les Vallettes, SSV. Forename **Jehannette**, 'little Jeanne'.
Jehans, la **Croix**	site: le Camp d'Ebat, T. wayside cross, probably relating to forename (m) **Je**(**h**)**an**, 'Jean's cross'. croix q.v.
Jehans, les	district, fields: S of le Colombier, T. Extant surname **Jehan** derived from forename (m) spelt **Jean**, **Johan**, **John**.
Jeffrey, de	see **Giffré**.
Jeigne(**s**), les	see **Jaigne**.
Jempreys, les	fields: les Osmonds, SSP. Recorded as 'courtils **Jean Prey'**, 1719.
Jenas, la **Vallée**	house, land: le Monnaie, les Niaux, SA. Surname extinct 16th c. **Du Genât**. 'Mr Genât's' valley.
Jenâ(**t**)**s**, les	see **Genâts**.
Jénémie, la **Vallée**	land: la Cache, SPB. Surname extinct 17th c. **Jénémie**. Also spelt Jéné**mye**, **Genemye**. croix q.v.
Jénémies, les	district: les Jénémies, SSV. Derivation as above.
Jénétier, le	land, fields: le Neuf Chemin, la Vieille Rue, SSV; les Raies, SPB; la Mare, Pleinmont, T. 'patches' of broom (genet). Also spelt **Jénéquet**.
Jenêts, à	see **Genêt**.
Jenêts, le **Clos** à	field: les Annevilles, SSV. 'broom'.
Jenette, de	field: les Simons, T. Contraction of forename **Je**(**h**)**annette**.
Jénotières, les	land: Jerbourg, SM. Area of broom (genêt).
Jentez,	see **Gentez**.

Jerbourg, de	château (castle), peninsula: district S of le Mur (Doyle Column), historically part of seigneurie of Fief de Sausmarez, SM. Château, Doyle column, mur q.v.
Jerbourg, route de	road: to Jerbourg, le **Grand Chemin**, part of chemin le Roi q.v., skirting the island.
Jérémie, de	field: la Villiaze, SA. Recorded since 1843. Forename (m) or extinct surname, **Jérémie** 19th c.
Jersey, de	fields: le Carrefour, St George, C; le Clos Vivier, les Jetteries, SSV; les Laurens, SPB; le Grée, T. Extant surname **De Jersey**.
Jersiais, **Barnabé** Le	see **Barnabé** (Le Gersiez).
Jersiez, **Camp** au	land: les Mielles, VdelE.
Jet, le	fields: le Clos Hoguet (le Jet), SSV; les Norgiots (le Get), SA. Extinct surname **Le Geyt**.
Jethou, **de**	see **Gehou**, **De**
Jetteries, les	house, fields: les Massies, SSV. Derivation as above, **Le Geyt**
Job, de	field: Baugy, CduV. Forename (m) **Job**. First ref. 1755.
Job, **Fosse**	land: les Pelleys, C.
Joine Denis, de	field: Candie, C. from **Jehanne Denis** (spelt **Jouanne Denis**, 1700). Jehanne q.v.
Joinet, le **Camp**	house, land: le Four Cabot, SA. 'Jehannet's land'. Recorded **Camp Jehanet**, 1610; **Camp Jouanet**, 1930. Jehannet q.v.
Joli, de	fields: les Traudes, SM. Probably extinct surname **Tolley**. Spelt **Tolly**, 1667.
Jonc, le **Clos** au	field: Saumarez, C. 'common bullrush'.
Jonctiaux, les	field: le Truchon, SPB. 'rush patch'. Also spelt **Joncquière**.
Jonquet, le	field: les Arguilliers, VdelE; la Couture, SPB. 'small' rush patch.

Jonquerette, la — land: Pleinmont, T. Area of small growing rush.

Josué, de — former field: la Grande Rue, SM. After **Josué Maindonnal**, owner, 1667.

Jouanne, de — fields: le Douit du Moulin, SPB; la Carrière, F. Either forename **Jehanne** q.v. or surname extinct 18th c. (extant Jersey) **Jouanne**.

Jouanne, la croûte d'**Anne** — see **Damun Gouanne**, la **Croix**. croix q.v.

Jouans, les — field: la Villiaze, SA. From forename **Jehan**.

Jo(**u**)**bres**, de — field: Farras, F. Origin unknown, first recorded 1857.

Joullot, de — field: near les Landes du Marché, VdelE. Origin unknown.

Joupe(**s**), de(des) — land, field: la Mare (Hure ès Jouppes), les Laurens (les Joupes), T. Extinct surname **Joupe**.

Jourdain David, le **Bordage** — see **Davids**, les.

Jour Denis, de — field: les Huriaux, SM. Origin unknown.

Juas, la **Hougue** — area: les Juqueurs, CduV. Misspelling of la **Hougue Cas**, from **quas** in 16th c. script (1591). Cas q.v.

Julien, de — fields: CduV. Origin unknown. Recorded **Jelien**, 1591.

Julien, St — house: rest house or hôpital with chapel attached at bottom of St Julian's Avenue, SPP. Founded 1535 by Thomas le Marquant and wife.

Julienne, rue — road: la Rivière, C. Probably from extinct surname **Julienne**.

Juliennes, les — fields: le Frie, SPB. probably from above extinct surname **Julienne**. Recorded les **Geliennes**, 1671.

Juqueurs, les — fields: Bordeaux, CduV. Extinct surname **Le Juqueur**.

K

King's Mills — see **Moulin**, le **Grand**.

L

Labour, à — fields: la Bertozerie, SPP; les Portelettes, T; Jerbourg, SM. Arable fields reclaimed in 17th c. from open field strip farming.

Labour, **Camps** à — land: near le Mur Jerbourg. SM.

Lacheur, au — fields: les Beaucamps, C; la Forge, les Jehans (Israel Le Lacheur), T; les Villets (Jean Le Lacheur), F; la Fallaize, SM. Extant surname **Le Lacheur**. Douit d'Israel Le Lacheur q.v.

Lague, la — fields: E of le Catioroc, SSV; la Lague Hervy, SPB/T. Denotes a flat and heavy sandy beach. Hervy q.v.

Lain, à — fields: les Huriaux, les Prevosts, SSV. Origin unknown.

Laitte, de — fields: rue de Laitte, SPB/T. Surname extinct 15th c. **De Leythe**. Also recorded as **Leythe**, 1595.

Lamballe, de — see **Amballes**.

Lambert, de — field: la Hougue du Pommier, C. Surname locally extinct 17th c. **Lambert**.

Landais, le **Clos** — fields: les Clos Landais, SSV. 'De Lande's enclosure'. Surname extinct 15th c. **De Lande**. Accentuated as place name. Recorded **Clos Landez**, 17th c.

Lande Boutefèves, la	fields: les Boutefèves, SM. Boutefèves, q.v.
Lande(s), la(les)	houses, fields: les Longs Camps, les Salines, SSP; les Beaucettes, rue des Landes, CduV; Albecq, le Friquet, C; le Briquet, SSV; les Laurens, T; route des Landes, F; la Quevillette, les Huriaux, SM; le Hurel, SA. 'heathland'.
Lande, la **Grande**	fields: les Lohiers, SSV. 'the big heath'.
Lande, le **Clos Allis de**	land: la Hougue Patris, CduV. Enclosure late of Allis De Lande. Surname extinct 15th c. **De Lande**. Landais, le Clos, q.v.
Landelle, le **Camp Thomas De La**	land: le Désert, SSV. Derivation as above. camp q.v.
Landelles, rue des	road: N of les Effards, C.
Lande Martel, le	field: Farras, F. Martel q.v.
Landes, **Commune** des	land: rue des Landes, CduV. Former common land sold and enclosed in 1864.
Landes de la Foret, les	land, fields: originally N part of parish, F. Moulin à vent q.v.
Landes de la Foret, **Douit** des	stream: le Marais Gouis, F.
Landes du Marché, les	site: former expanse of heathland (area c. 200 vergées), VdelE. Medieval island market, central to the island, traversed by routes from all directions including le chemin le Roi. Moved to SPP by early 14th c. Roi, le Camp du, q.v.
Landes Hue, les	fields: W of les Laurens, T. 'Hue's heathland'. From surname **Hue** q.v.
Landes Priaulx, les	field: in Airport, F. Priaulx q.v.
Landes, rue des	road: formerly rue de Bémont, now called Donkey Hill, SM; W of le Bourg, F.
Landes Yvelin, les	fields: rue des Landes, SPB. Yvelin q.v.
Landry, la **Fosse**	former fields: la Fosse André, SPP. Surname

	extinct 14th c. **Landry**. Misspelt **Fosse André**, 18th c. Bordage, fosse q.v.
Landry, le **Bordage**	feudal holding of land, le Vauquiedor, SPP.
Langlois, de **Colin**	fields: les Monts Héraults, SPB. Extant surname **Langlois**.
Langlois, de **Jean**	fields: S of les Landes Hue, T. Derivation as above.
Langlois, de **Rachel**	field: le Clos Vivier, SSV.
Langlois, de **Rougier**	field: les Eperons, SA. Naming dates from 17th c.
Largisse, la	land: les Hougues, la Hurette, C. 'verge by road side'.
Larron, la **Fosse** à	cove: les Pézérils, T. 'concealed inlet'.
Larron, le **Petit**	land: La Crève Cœur, CduV. 'hidden'.
Laurens, les	district: E part of parish, T. Forename (m) corresponds to forename (f) Lorenche. Recorded as placename from 1705, earlier as late of **Lorens Gallienne** of les Galliennes, T, 1595. Spelling altered from Lorens to Laurens in 19th c. Lorens q.v.
Laurens, la rue du **Courtil**	road: N of rue de Laitte, la vieille rue, les Galliennes T.
Laurens, le **Courtil**	field: les Buttes, T.
Laurens, le **Vau**	land: St Julian's avenue, SPP. Forename (m) Lorens. val q.v.

Le Surnames with an initial **Le**, used as proper names for part of a place name, are indexed under the substantive name.

Le **Beir**, Le **Baillif**, Le **Beuf** (Le **Boeuf**), Le **Blanc**, Le **Bouteillier**, Le **Breton**, Le **Canely**, Le **Canu**, Le **Cauchez**, Le **Cauf**, Le **Cherf**, Le **Chevallier**, Le **Cocq**, Le **Comte**, Le **Cornus**, Le **Coutanchez**, Le **Couteur**, Le **Cras**, Le **Cucu**, Le **Cucuel**, Le **Fauconnaire**, Le **Francez**, Le **Febvre**, Le **Feyvre**, Le **Gallez**, Le **Gelley** (Le **Gelé**), Le **Gersiez**, Le **Geyt**, le

	Gois, Le **Gouais**, Le **Goubey**, Le **Huby**, Le **Huray**, Le **Jersiez** (Le **Jersiais**), Le **Juquier**, Le **Lacheur**, Le **Lièvre**, Le **Maitre**, Le **Marinel**, Le **Messurier**, Le **Mierre**, Le **Moigne**, Le **Page**, Le **Pelley**, Le **Prestre**, Le **Rouf**, Le **Roy** (Le **Roi**), Le **Sage**.
Lébergeade, de	field: les Adams, SPB. Origin unknown. **L'Hebergarde**, 1671. **Lébergade**, 1696.
Lefebvre, rue	road: Constables Office, SPP. W side of High Street, near le Carrefour.
Legat, Fief au	feudal holding of land, now absorbed into Fief St Michel. VdelE.
Lenfestey, de	field: S of la Grande Mare, C. First recorded 1853.
Lengage Faras, de	see **Farras**.
Lengier, le **Clos**	field: S of la Grande Mare, C. Surname extinct 15th c. **Lengier**.
Leonard Paré, de	see **Paré**.
Lerée, de	see **Erée**, l'.
Lesant, la **Terre**	land: la Couture, SPP.
Lesant, le **Bordage**	feudal holding of land, N of la Mare Pirouin (Bouet) SPP. Surname extinct 15th c. **Lesant**. Also feudal holding of land, Fief **Lesant**, SM.
L'Espesse, Fief de	feudal holding of land, now known as Fief de Beuval. SPB.
Lestournelerie, la	see **Etonnelerie**.
Lestrainfer, de	see **Estrainfer**.
Letat, la **Hure** de	see **Crolier**, la **Hure** du, T.
Letonnellerie, la	see **Etonnelerie**.
Letournel, de	see **Etonnelerie**.
Leur, la	see **Alleur**.
Levin, Fief de	feudal holding of land. SM.

Lévre, à la	field: la Croute Becrel, CduV. Origin unknown. à la **Laiure**, 1591; à la **Leure**, 1654.
Lichoteries, les	house, fields: Rocquaine, SPB. Surname extinct 15th c. **Le Lichotel** (**John Le Lichotel**, 1483).
Lièvre, le **Carrefour** au	cross roads: Fermains, SPP. Extant surname **Le Lièvre**.
Lif, la **Hure** de	land: near la Prévôté, SPB.
Lif, le **Val** de	fields: la Prévôté, SPB. 'lily'.
Ligneris, les	fields: Havilland (vale), SPP; le Braye, CduV; les Houmets, C; les Jénémies, SSV; Bon Port, Jerbourg, SM. 'alignments' Also spelt **Lignères**, **Lignères**.
Lihou, d'**Abraham**	field: la Pitouillette, C. Extant surname **Lihou**.
Lihou, de	house, land: les Reveaux, SPB. Named in early 19th c. when built, because of a view of Lihou Island. But also called 'le Neuve Maison du Long Frie'.
Lihou, de **Calistre**	field: Candie, C. Extant surname **Lihou**.
Lihou, de **Guillaume**	field: les Niaux, C. Derivation as above.
Lihou, de **Jean**	field: le Belle, SPB. Derivation as above.
Lihou, Fief de	see **Lihou**, fief du prieur de.
Lihou, fief du prieur de	feudal holding of land, Fief de Lihou, VdelE; SSV; SPB; T.
Lihou, (ile de)	island: tidal, off L'Erée peninsula, SPB. Also extant surname **Lihou**. 'hou' denotes island.
Lihou, le **Braye** de	see **Braye**.
Lihou, le **prieurie** de	priory of **Notre Dame de Lihou**, formerly subsidiary to le **prieurié du Valle** q.v. Alienated to the crown by the early 15th c. and abandoned and ruined before the Reformation.

Lihoumel, le	islet: tidal, W of Lihou island.
Lisle, de **Jean De**	field: la Mare, SPB. Extant surname **De Lisle**.
Lisle, de **Mes** (**Mr**) **De**	field: la Hougue Fouque, SSV . Derivation as above.
Lisle, de **Massy De**	house, land: le Tertre, SA. Derivation as above.
Lisle, de **Nicolas De**	fields: les Islets, les Raies, SPB. Derivation as above.
Lisle, de **Thomasse**	field: (unlocated), SA. Derivation as above.
Lisles, les **De**	houses, fields: St George (les Delisles), C; les Fontenelles, SSV. Derivation as above.
Lobi, **Lobis**, **Loby**	see **Obis**.
Loc, **Fosse** à la	land: near the King's Mills, C.
Loe, **La**	fields: les Goddards, les Grands Moulins, la Rivière, C. Surname extinct 16th c. **La Loe** (**La Lowe**, 1309).
Loges Cocquerel, les	field: Jerbourg, SM. '**Cocquerel's** huts'. Cocquerel q.v.
Lohiers, les	district, fields: les Lohiers, SSV; les Adams, les Sages, SPB. Surname extinct 16th c. **Lohier**.
Loisel, à	see **Oisel**.

Long (**m**), **longue** (**f**) with proper names listed by parish, see also individual entries.

SPP	Longs **Camps**, Longue **Raye**, Longue **Rue**, Longue **Rocque**.
SSP	Longs **Camps**, Longue **Hougue**, Longue **Rocque**.
CduV	Long **Bourel**, Longr**ée** (Longue **Raye**), Longue **Chapelle**, Longue **Rocque**.
VdelE	Longue **Giolle**.
C	Long **Désert**, Long **Port**, Longue **Pierre**, Longue **Pièce**.
SSV	Long **Bos**, Longs **Camps**, Longue **Pièce**, Longue **Pierre**, Longue **Raie**, Longue **Rue**.

SPB	Long **Avaleur**, Longs **Camps**, Long **Frie**, Long **Mare**, Longue **Pierre**, Longue **Rocque**.
T	Long **Avaleur**, Longue **Pièce**.
F	Longs **Camps**.
SM	Longs **Camps**, Long **Trac**, Longue **Rue**.
SA	Longs **Camps**, Longue **Pièce**, Long **Pierre**.
Longbeaux, les	see **Bos**, les **Longs**.
Longrais, de	see **Raie**, le **Long**.
Longraie, le	see **Raie**, le **Long**.
Long Raie, la **Croûte** dite	land: la Haute Banque, CduV.
Longrée, de	fields: l'Ancresse, North side, CduV. 'long furrow'. Recorded **Longue Raye**, 1591. Raie, q.v.
Longs Camps, **Douit** des	land: les Long Camps, SPP.
Longs Camps, la **Houguette** des	land: les Long Camps, SSP.
Longs Camps, rue des	road: N of les Bas Courtils, SSV.
Longstore, the	artillery store: le Bouillon, SPP. 18th c. building.
Longue, la **Cour** de	see **Cour de Longues**.
Longue Raye, **Camp** à la	see **Raie**, le **Longue**
Longues, Fief des	feudal holding of land. Dependency of Fief le Comte. SSV.
Lorenche(**s**), de	fields: les Grantez, C; la Roberge, F. Forename (f) **Lorenche**. Corresponds to Lorens (m) q.v.
Lorenche, la **Hougue**	land: near les Vardes, SSP.
Lorens, de	fields: les Buttes, T. Forename (m) Lorens, Laurens, q.v. Recorded late of **Lorens Gallienne**, 1595.
Lorette, la **Chapelle** de la	site: to N-E of Victoria Tower (moulin à vent de l'Hyvreuse), SPP. Named after

	Santa Casa di Loreto (the Holy House), in Italy. Chapel, chapelle q.v.
Lorier(**s**), le(s)	houses, fields: Croix du Bois, les Mielles, CduV; on parish boundary between SPB and SSV, to E of la Houguette school. Surname extinct 18th c. **Lorier**, but now extant following resettlement.
Lorreur, du	land: les Landes, C. Surname extinct 15th c. **Le Lorreur**.
Lot, le **Grand**	land: le Friquet, C. 'lot', (i.e. 1 vergée 20 perches)
Lot(**t**)**ais**, les	fields: le Gron, SSV; les Landes, F; in Airport, SA. 'lotted'.
Loué, de **Trop**	see **Trop Loué**.
Louis, de	fields: la Pomare, la Cloture, SPB. Forename (m) **Louis**.
Louis, la **rue** de	road: W of les Lichoteries, SPB. Derivation as above.
Lubin, le	field: Airport, F. After **Lubin Falleze**, 1501.
Lucasée(**s**), les	fields: Fermains, SPP; les Landes, CduV. Forename (m) or surname, **Lucas**.
Lucas Arnault, Fief de	feudal holding of land. C. Always perched with Fief Dom Jean le Moigne, q.v.
Lucas, de	field: les Capelles, SSP.
Lucas, le **Clos**	land: unlocated. SA.
Lucas, les **Camps**	field: les Villets, F.
Luce, le **Val**	field: SW of le Vallon, SM. Extant surname, **Luce**.
Luet, le **Moulin**	mill: former watermill (moulin à eau), Moulin Huet, SM. Surname extinct 15th c. **Luet**. Recorded as **Huet** 18th c. [Capital L in older script is easily read as H.]
Luet, de	fields: Bon Port, SM. Derivation as above.

M

Maban, **Hougue**	land: unlocated. SSV.
Mabelle, de	field: rue a l'Or, SSV. Origin unknown. Earliest record **'Mabel'**, 1616–1854.
Mabire, le **Clos**	field: SPB. Extant surname, **Mabire**. clos q.v.
Machon, de	land: harbour, route Carre, SSP; les Ruettes Brayes, SM. Extant surname **Machon**.
Machon, le **Val** au	field: la Neuve Maison, SSV. Derivation as above.
Madeleine, la	houses, land: S of les Islets, SPB, after **Magdeleine Gallienne**, widow of Abraham Lenfestey, early 18th c.
Mado, de	field: les Traudes, SM. Surname extinct 17th c. **Mado**.
Mado, le **Fosse**	land: les Hubits, SM.
Mado, le **Mont**	house, land: les Hubits, SM. Derivation as above. hurel, mont q.v.
Magloire, **Saint**	site: St Magloire, CduV. Former chapel. Recorded, medieval legend, **St Magloire**. Also recorded as **St Maillère**, 1591, 1654, **St Magloire** 1727 to date. Maillière, q.v.
Magues, les **Hougues**	house, land: W of les Effards, SSP. Extinct surname **Mague**.
Mahaut, de	houses, land, road: Collings road, les Rohais, SPP . Surname extinct 17th c. **Mahaut**.
Mahaut, la rue	roads: Hacsé, CduV; N of Richmond, SSV. Derivation as above. rue q.v.
Mahie, le **Creux**	hollow: les Tielles, T.' deep cove'. Medieval forename (m) and surname. creux q.v.
Mahiel, le	house, field: les Prevosts, SSV. after **Thomas Mahiel**, (contract, 1676).

Mahiette, la	fields: les Cambres, les Tielles, T. Diminutive of **Mahie**, with reference to land.
Mahy, de	former fields: le Bouet, SPP; les Annevilles, VdelE. Extant surname **Mahy** (Belgian immigrant, early 16th c.).
Maillère, **Saint**	see **Saint Magloire** (possibly St Maiolus, d. 994, Benedictine abbot of Cluny).
Maindonal, de	field: la Varde, SPP. Extant surname **Maindonal**.
Maindonnaux, les	district: SM. Derivation as above.
Maindonnaux, rue des	road: les Maindonnaux, SM. Derivation as above.
Maingy(**s**), les	house, field, land: Belval, CduV; Cauchebrée, VdelE. Surname extinct 20th c. **Maingy**, **Mainguy**. Place name dates from 18th c.
Maingy, la **Croûte**	les Maingys: VdelE. Derivation as above. croûte q.v.
Maingy, la rue	les Maingys: VdelE. Derivation as above. rue q.v.
Main Moulin, le	see **Moulin à eau**, le, les **Grands Moulins**.
Mainnel, le	field: CduV. Origin unknown, possible connection with **Meunier** q.v. Earliest record **Mainnier**, 1591, 1654.
Mains, de **James**	field: les Effards, SSP. Earliest record 1878. Extinct surname **Mains**. les Monmains, q.v.
Mains, **Grand**	land: unlocated, SSP.
Maisière, la	see **Messière**.
Maison(**s**), la(les)	houses, all parishes. Below are listed houses with descriptive and proper names. See also individual entries.
Maison Bonamy, la	house: (now Ashbrook), les Quartiers, SSP. Surname extinct 19th c. **Bonamy**.

Maison Bonamy, la — house: la Pomare, SPB. After **Helier Bonamy**, 1768. Now a ruin.

Maison d'Aval, la — 'lower house, house on a slope',

1. house: Sohier, CduV. A field in 19th c.
2. house: rue d'Aval, VdelE. 15th – 16th c. house.
3. house: near les Prevosts, SSV. 15th – 16th c. house.
4. house: near la Cour de Longues, SSV.
5. house: les Messuriers, SPB. Former rectory, sold 1441 (still exists, joined to early 19th c. villa).
6. former house: below les Islets, SPB. Ruined.
7. house: la Bellée, T.
8. house: near les Brouards, T.
9. house: near les Houets, T.

Maison de **Bas**, la — 'bottom house';

1. houses: N of les Haizes, CduV.
2. house: formerly les Beaulins, SA. Beaulins q.v.

Maison de **Haut**, la — 'upper house';

1. house: at Maison au Comte, CduV.
2. house: les Grands Moulins, C.
3. house: St Germain, C.
4. house: S of les Padins, SSV.
5. house: les Messuriers, SPB.

Maison de **St Pierre**, la — former house: rue des Grandes Rues, SPB. Demolished 19th c. [replaced with cottage].

Maison de **St Sauveur**, la — former house: rue St Pierre, SSV. Demolished 19th c. [now a field]. St Sauveur q.v.

Maison des **Pauvres** — house: les Islets, SPB. Former poor house from bequest of Thomas De Lisle to poor of parish, early 17th c.

Maison Dogon, la — house, field: le Préel, C. Dogon q.v.

Maison Naftel, la — house: le Bouet, SPP. Demolished in 2001. Extant surname **Naftel**.

Maison, la **Grande**	'large' house; S of Maison d'Aval, VdelE.
Maison, la **Longue**	house: near le Neuf Clos, SSV. 'long' house, originally three attached dwellings.
Maison, la **Neuve**	'new house'. 1. house: near le Villocq, C. Built late 18th c. 2. houses, district: SSV. Built 16th – 17th c. 3. field: les Paysans, SPB. No record of a house since 1671. 4. house: at les Vallées, T. Built early 19th c. [understood to be recently renamed].
Maison, la **Rouge**	house: at la Hougue Fouque, SSV. Stonework concealed by pink pebble dash render. House rebuilt 1990s in random stone.
Maison, la **Vieille**	house: le Hurel CduV. 'old house'.
Maison, la **Vingt**	land: le Friteaux, SM. Origin unknown, first recorded 1654, field 1667.
Maisons au **Comte**, les	houses: W of Baugy, CduV. Extant surname **Le Comte**, **Le Conte**.
Maisons Brulées, les	road, ruette: off la Haut Pavé (Mill Street), SPP. Presumably a medieval disaster. Burnt Lane, Port(e) Vase q.v.
Maisons de **Garde**, les	see **Garde**.
Maisons, **Grandes**	houses: les Grandmaisons, SSP.
Maisons, les **Vingt**	field: le Variouf, F. Origin unknown, recorded as field in 1671.
Maitre Guillaume, de	house, field: la Boullerie, SA. Probably pre-Reformation. Recorded 1610.
Maitre, au	fields: Fort George, SPP; le Friquet, C. Extant surname **Le Maitre**.
Maitre, de **Jean Le**	field: les Cambrées, T. Derivation as above.
Maitre, de **Pierre Le**	field: les Rocques, T. Derivation as above.
Maitre, **Hougue** au	land: unlocated. CduV.

Malade, le **Vaux** — field: S-W of les Villets, F. Possibly connected with Maladerie (see below). val q.v.

Maladerie, la —
1. land: E of les Terres, SPP.
2. area: off Sandy lane, SSP.
3. area: W of Fort Richmond, SSV.
4. area: W of les Huriaux, SM.

In four isolated areas, possibly 13th and 14th c. 'leprosaria'. Each area seems to be about 3-4 vergées, and may incorporate burial pits for a number of 'Black Death' victims, mid 14th c. (See Current Archaeology, Nov. 2009 p.236.)

Malassis, le — fields: Fermains, SPP; les Huriaux, SM; W of la Brigade, SA. 'troublesome, difficult'.

Mal Etant, le — see **Etang**.

Malherbes, les — field: les Laurens, T. First recorded 1896. 'weeds'.

Mallard, le — fields: la Villiaze, F. Surname extinct 15th c. **Mallard**.

Malpeines, les — fields: le Passeur, CduV; les Vallées, N of Saumarez park C; les Laurens, T. 'lacking vigour'. Recorded **Malle Payne**, 1549, **Malles Poinnes**, 1591. Also spelt **Malpoignes**.

Manchelle, la — land: South Side, SSP. Origin unknown.

Manchot, le — field: les Domaines, SSV. Origin unknown. Also spelt **Mancheaux**.

Manchot, le **Douit** — stream: near le Fontaine, SSV.

Manchotte, la — field: les Crabes, SSV. Origin unknown.

Manel, le — fields: rectory, SPB; les Coutures, Sausmarez manor, la Colombelle, SM; le Monnaie, SA. 'handful'.

Manoir)s), le(s) — medieval manors :

1. **Franc Manoir au Marchant**, Lefebvre street, SPP. Replacement buildings on site.

2. **Manoir de la Grange**, La Grange, SPP. Site lost.
3. **Manoir de Havilland**, les Vardes, SPP. Site lost.
4. **Manoir d'Anneville**, SSP. Built by Sir William de Chesney, 1250–1260s. Reconstructed 1980s.
5. **Manoir de Ste Anne**, Les Grands Moulins, C. Within Fief St Michel, site lost.
6. **Manoir de la Perelle**, SSV. Possibly house now called 'les Grands Courtils'. chapelle q.v.

Merchant manor houses – mansions dating from the 15th and 16th centuries, include:

1. **Granges de Beauvoir** les Rohais, SPP.
2. **Normanville** Fosse Andre, SPP.
3. **Vrangue Manor** la Vrangue, SPP.
4. **St George** les Delisles, C.
5. le **Manoir** New road, F.
6. **Sausmarez Manor** route de Sausmarez, SM.
7. les **Rohais** les Rohais de haut, SA.

Manquées, les — field: les Massies, SSV. Surname extinct 17th c. **Le Manquais**.

Manquin, le **Val** — land, valley: E of les Tielles, T. Extinct surname, spelt **Mannequin**, 1595. val, fonderille, q.v.

Mansell, de — land, road: Mansell street, contrée croix Mansell, SPP; rue des Marais, VdelE; les Cornus, SM. Extant surname **Mansell**. Also spelt **Mauncez**, 16th c. croix q.v.

Mansell, le **Clos** — field: le Mont Saint, SSV. Derivation as above. clos q.v.

Marais, le(s) — houses, land, fields, all parishes. 'marsh'. Often a decayed natural pond. Earlier spellings **Maresc**, **Marescq**. mare, planque q.v.

Marais Dallebec, le — marsh: 13th c. name for le **Grand Marais**, [la **Grande Mare**]. Ref. 16th c. **Maresc Dalebec**. Albecq q.v.

Marais d'Orgeuil, le — fields: district of le Château des Marais, SSP. 11th to 14th c. name for area. Orgeuil q.v.

Marais du **Bullaban**, le — area: large area N of rue des Marais, VdelE. Bullaban q.v.

Marais, **Château** des — castle: medieval castle of refuge on 'hougue' in large area of marsh, le **Marais d'Orgeuil**. château, orgeuil, pont q.v.

Marais Gouie, rue du — road: unlocated, SPB.

Marais Gouies, les — fields: E of Plaisance, SSV; SPB. Former marsh. Gouies q.v.

Marais, le **Bas** — fields: N of Hougue du Moulin, CduV; Perelle, SSV. 'low lying' marsh.

Marais, le **Grand** — see **Mare**, le **Grand**.

Marais, le **Petit** — fields: W of la Grande Mare, C. Drained and enclosed, 17th c.

Marais, le **Pont** du **Petit** — former crossing: Braye du Valle (midway). pont q.v.

Marais, le **Rond** — enclosures: l'Ancresse, CduV. Drained and enclosed, 18th c. Let as Crown fee farm, 'My Lady', since 1739.

Marais, rue du — road: W of les Rouvets, VdelE.

Marais Ysorey, le — land, fields: N of les Cherfs, C. Former marsh. Reference 16th c. spelt **Marescq Ysorey**. Orais q.v.

Maraitaine, la — house, land: Baugy, CduV. 'marshy' land.

Maraive, la — house, fields: W of la Lande, CduV. 'deep marsh'. Recorded **Mareyve**, 1591.

Marchant(**s**), le(s) — land, fields: les Guelles, SPP; Ste Anne, les Touillets, Seaview Farm, C; la Hougue Anthan, SPB; les Blicqs, SA. Extant surname

Le Marchant. Also feudal holding of land, Fief au **Marchant**, SM.

Marchant, **Le**, **Fort** — fort: W end of Fontenelle Bay, CduV.

Marchants, **Grands** — land: Seaview Farm, C.

Marchants, les **Camps** — land: l'Epinel, F. Derivation as above. camp q.v.

Marche, de — fields: rue à l'Or, la Ville au Roi, SPP; la Hougue Patris, CduV. Surname extinct 19th c. **Marche**, **Marcherie**. Marchez q.v.

Marche, la **Grande** — road: now 'King's Road', SPP.

Marche, la **Petite** — road: now 'The Queen's Road', SPP. Part of le chemin le Roi. 'Superior roadway' parallel to la Grande Marche (q.v.) now the 'King's Road (q.v.) Both of equal length with roads at N and S to complete a quadrilateral.

Marché, le — the market;

1. les Landes du Marché, VdelE. medieval market.
2. la Fontaine Cache Vassal on N, to le Val Videcocq on S, SPP. Late medieval period, le chemin le Roi going through the middle.
3. Vicinity of Town Church, SPP, in early modern period, followed by the 19th c. market buildings.

Marcherie, la — house, land: la Villette, SM. Derived from surname extinct 19th c. **Marche**.

Marcherie, la — house, land: Saints, SM. Demolished during the German Occupation. Derivation as above.

Marchez, les — district: N of les Bruliaux, SPB. Derivation as above.

Marchez, la **Croix** des — district: les Marchez, SPB. Former wayside cross. Derivation as above. carrefour, croix q.v.

Mare(s), la(les)	houses, land, fields, all parishes, natural pond.
Mare de **Carteret**, route de la	road: N-W of la Mare de Carteret, C.
Mare, la **Claire**	former pond: N of le Felconte, ½ SSV, ½ SPB, bounded by road on S and E, by track on W. SSV held by Fief le Comte and SPB by abbot of Mont St Michel. Presumably a 'clear' pond, noted for an eel fishery in 17th c. prior to drainage in 18th c.
Mare, la **Claire**	house, land: to S-W of above pond.
Mare, la **Grande**	formerly le Grand Marais, C. 'large' marshy pond. Drained and enclosed as Crown fee farm 1739.
Mare, la **Hure**	buildings, land: N of North side, CduV. 'head of growth' on pond.
Mare, la **Hure** de la	land: on top of Pleinmont, T.
Mare, la **Longue**	fields: W of le Catillon, SPB. Former 'long' pond.
Mare, la **Plate**	area: l'Ancresse, CduV. Former 'shallow' pond.
Mare, la **Rousse**	land: E of l'Erée, SPB, 'reddish stained soil', noted fishery 14th & 15th c. Drained and let as fee farm since 1739.
Mare, rue de la **Claire**	road: N of rue du Felcomte, SPB.

Mare(s) – former ponds, with proper names, are listed below. See also individual entries.

Mare à **Pleinmont**, la	la Mare, Pleinmont, T.
Mare au **Chanteur**, le	la Rousaillerie, SPP.
Mare au **Vecque**, la	les Pièces, F.
Mare au **Vicheulx**, la	N of rue Rocheuse, SPB.
Mare Bazille, la	la Hougue à la Perre, SPP.
Mare Côbo la	le Mont Crevel, CduV.

Mare de **Carteret**, la	N of Côbo, C.
Mare de **Nermont**, la	les Barrats, VdelE. (**Mare de Noirmont**).
Mare De Putron, la	Fort George, SPP.
Mare de **Rocque Poisson** (**Paysant**), la	S of rue Rocheuse, SPB.
Mare de **St André**, la	N of les Vauxbelets, SA.
Mare Denis, la	les Douvres, SM. (**Mare au filx Denys**, 1541).
Mare ès **Anges**, la	les Mares, SPB. 'monkfish'.
Mare ès **Mauves**, la	l'Ancresse, CduV. Modern name.
Mare ès **Querpentiers**, la	S of Hacsé, CduV.
Mare Hailla, la	les Salines, SPB. (**Mare Harla**).
Mare Mado, la	W of les Hubits, SM.
Mare Mauger, la	les Marais, SSP.
Mare Pelley, la	rue Colin, VdelE.
Mare Pirouin, la	le Bouet, SPP.
Mare Revel, la	S of le Douit du Moulin, SPB.
Mare Rougier, la	E of Côbo, C.
Mare Salliot, la	la Croûte au Sage, CduV.
Mare Samsonnet, la	Bordeaux, CduV.
Mare, de **Thomas De La**	field: W of les Capelles, SSP. Extant surname **De La Mare**.
Mares, **Croûte** des	land: unlocated, SPB.
Mares, de **Collas Des**	field: E of les Islets, SPB. Surname extinct 16th c **Des Mares**, also **Desmares**.
Mares Pellées, **Hougue** des	land: les Mares Pellées, CduV.
Mares Pellées, les	W of la Grange, CduV.

Maresq, le **Clos Du** — see **Dumaresq**.

Maresquet(s), le(s) — houses, fields: la Ronde Cheminée, Dobrée, SSP; la Hougue Jehannet, Bordeaux, CduV; E of les Maingys, VdelE; N-E of Saumarez park, N of les Goddards, C; le Vallet, SPB; N-W of le Tertre, SA. Diminutive of marais q.v. 'small marsh'.

Maresquet Piart, le — fields: la Rochelle, CduV. Piart q.v.

Maresquières, les — fields: W of le Rocher, VdelE; les Bas Marais, SSV; les Landes Hue, T. Also spelt **Marestières**, 'swamp'.

Marette, la — houses, fields: S of les Grandes Capelles, le Clos, SSP: Quatre Chemins, CduV; N of le Désert, N of Fort Richmond, les Lohiers, SSV; les Paysans, SPB; les Landes, F; rue des Marettes, SM. Diminutive of mare, q.v.

Marettes, rue des — see Marette.

Margion, la — houses, land: le Hamel, C. 'coastal margin'.

Margot, de — field: la Palloterie, SPB. Forename (f) **Margot**.

Marguerite, la rue — road: now called New street SPP. After **Marguerite**, wife of William Le Marchant, bailiff 1770–1800.

Marie, de — field: le Villocq, C. Forename (f) **Marie**.

Marie Girard, de — field: les Beaucamps, C.

Marie Le Feuvre, de — field: king Edward VII hospital, C.

Marie Moullin, de — field: Farras, F.

Marin, de **Collas** — field: les Maumarquis, SA. Surname extinct 17th c. **Marin**.

Marin, **Douit** du — field: near les Mauxmarquis, SA.

Marinde, la **Croûte** — fields: Oberlands, SM. Extinct surname **Marinde**.

Marinel, du — field: les Ozouets, SPP; house, le Marinel SPB. Surname extinct in Guernsey, 17th c. **Le Marinel**.

Marinel, le **Clos**	land: N of le Grantez, C.
Marion, la **Croûte**	fields: les Vardes, SSP; les Maumarquis, SA. Forename (f) **Marion**.
Marotte, de	field: Bordeaux, CduV.
Marottes, la rue des	road: la Hougue au Comte, leading to les Corneilles, C. 'steep, narrow road'.
Marquand(**s**), les	fields: le Neuf Chemin, SSV. Extant surname **Marquand**.
Marquand, la **Bouvée**	feudal holding of land. Dependency of Fief Gohiers, SSV.
Marquand, la rue	road: E of les Naftiaux, SA. Derivation as above.
Marquet, de	field: les Vallées, C. After **Marquet Drognet**, 1470.
Martel, de	fields: le Pont Renier, les Marais, SPB; les Buttes, SA. Extant surname **Martel**.
Martello towers	1. **Fort Grey**, replaced Château de Rocquaine. 2. **Fort Saumarez**, (l'Erée tower). 3. **Fort Hommet**. towers built between 1804 & 1806

(Pre) **Martello** towers. Originally 15 towers all built between 1778–1779.

No. 1 & No. 2,	la Hougue à la Perre, Belle Greve battery (both demolished), SPP.
No.3.	le Mont Crevelt, SSP.
Nos. 4, 5, 6, 7, 8, 9	l'Ancresse, CduV.
No.10	Chouet, CduV.
No.11	Rousse, VdelE.
No.12	Vazon, C.
No.13	Petit Bot, F.
No.14	Saints, SM.
No.15	Fermains, SPP.

Martin, de	fields: les Gigands, SSP; le Rocher, les Landes, rue Cohu, C; Airport, F; les

	Traudes, SM; les Pointes, SA. Forename (m) **Martin**.
Martin Guignon, de	see **Guignon**.
Martin Tardif, de	see **Tardif**.
Martin, de **Collas**	field: les Houmets, C.
Martin, de **Colin** (**Collas**)	field: les Blanches, SM.
Martin, de **Pierre**	field: Woodlands, C.
Martin, la **Croix**	site: former wayside cross. SSV/SPB boundary at shingle bank on coast. Forename (m) **Martin**. croix q.v.
Martins, les	houses: N of rue Sauvage, SSP; N of le Truchon, SPB; W of la Villette, SM. Extant surname **Martin**.
Martins, la **Croûte** des	land: les Martins, SSP.
Martins, **Hougue** des	land: les Martins, SSP.
Masse, (du **Moulin**)	artificial mound on which a wind mill (moulin à vent) was built.
Masse, la	1. field: W of la Grand Marche (Queen's Road), SPP. 2. house: land, la Masse, les Capelles, SSP, near site of moulin à vent of Fief d'Anneville. 3. field: les Loriers, CduV. 4. house: la Masse, C. Adjacent to old school building, Castel School. Site of 'masse' of moulin à vent of Fief le Comte. 5. field: la Colombelle, SM.
Massies, les	house, land: les Massies, SSV. Heirs **Thomas Massy**, 1662.
Massiotte, la	field: Farras, F. 'small mound'. mazotte q.v.
Massy, de	fields: la Gallie, SPB; la Corbière, F; les Hunguets, SA. Forename (m) or extinct surname **Massy**.

Massy, de **Gervais**	field: rue au Pretre, C. After **Gervais Massy**, early 16th c.
Massy, le **Camp**	land: le Haut Chemin, SPB.
Massy, le **Clos**	field: les Prevosts, SSV. After **Massy Le Dien** (Dyang), late 16th c.
Massy Brehaut, le **Camp**	see **Brehaut**.
Massy Gros, Fief	see **Gros**.
Massy De Lisle, de	see **Lisle**, **de**.
Masure, la	land: les Abreuveurs, SSP; les Landes, CduV; Saumarez park, C; les Jetteries, les Annevilles, SSV; les Adams, la Vallée, SPB. 'ancient ruin'.
Matheline, les **Messières**	field: la Hougue Anthan, SPB. Extinct surname **Matheline**. mèssieres q.v.
Matthieu, de	fields: la Madeleine, la Pomare, SPB. Forename (m) **Matthieu**.
Mauclos, le	field: facing house at le Tertre, C. 'poor enclosure'. Also called **la Houguette au Tillier** q.v.
Mauconvenants, Fief	feudal holding of land. Dependency of Fief de Longues, SSV.
Mauger, de	fields: les Baissieres, VdelE; les Jaonnets, SSV; les Fontenelles, F; les Blanches Pierres, SM. Extant surname **Mauger**, formerly **Maugeur**.
Mauger, la **Croûte**	field: le Hamel, VdelE. croûte q.v.
Maumarquis, les	houses, district: les Maumarquis, SA. Surname extinct 15th c. **Malmarchy**. Also feudal holding of land, Fief des **Maumarquis**, SA.
Maupas	see **Maurepas**.
Maupertuis, de	land, field: top of le Truchot, SPP; Marais du Bullaban, VdelE. 'bad hole', 'badly drained'.

Maupertuis, rue de	road: former name for the Truchot, SPP.
Maurepas, la rue du	road: Normanville, SPP. Origin Maupas, 'bad pathway'. rue q.v.
Mauvaret, la **Croûte**	fields: St Magloire, CduV. Origin unknown. croûte q.v.
Maze, le **Clos**	field: la Villette, SM. Extinct surname **Maze** (**Masse**).
Maze, la rue	road: W of la Grande Rue, SM. Derivation as above.
Mazotte, la	houses, land: les Landes, CduV. 'small mound'. Spelt **Massotte**, **Masete**, **Masote**, 1591. massiotte q.v.
Mèche, à	fields: les Nicolles, SSP; la Turquie, CduV; (unlocated) VdelE; les Girards, les Bergers, C; la Grande Rue, les Annevilles, SSV; les Martins, SPB. 'dryings' for wicks, plaits. Also spelt **Mecques**.
Mèches, **Clos** au	land: near the Castel Hospital, C.
Megaliths	1. standing stones, see **Chaise**, **Perron**, **Pierre**, **Rocque**. 2. chambered tombs, see **Castel**, **Creux**, **Déhus**, **Déhuzel**, **Pouquelaie**, **Trappe**, **Trépied**.
Melin, le	see **Meslin**.
Meltier, le **Pont**	field: les Messuriers, SSV. Surname extinct 15th c. **Le Melletier**. pont q.v.
Ménage (m)	enclosure or field, outside the arable and usually part of the 'curtilage' of a property.
Ménage(s), le(s)	houses, fields, land, all parishes. Also recorded as; le **Grand Ménage** – large le **Petit Ménage** – small le **Vieil Ménage** – old (established) le **Ménage d'Aval** – lower

	le **Ménage de Bas** – below le **Ménage de Haut –** upper
Ménage Bazin, le	field: le Rocré, C. Bazin q.v.
Ménage Corbin, le	fields: les Adams, SPB. Corbin q.v.
Ménage du Four, le	field: les Issues, SSV. Four, Du q.v.
Menage, **Grand**	land: le Chene, les Houards, F; le Briquet, les Frances, les Domaines, SSV.
Ménage Heaume, le	fields: les Villets, les Nicolles, F; SM. Heaume q.v.
Ménage Tourgis, le	field: la Neuve Maison, SSV. Tourgis q.v.
Menages, **Douit** des	stream: les Menages, Bon Port, SM.
Menages, **Fontaine**	spring: Bon Port, SM.
Mer, la	fields, cliffs: N of Fermains, SPP. 'the sea'.
Mer, **Camps** de la	see **Mer**.
Mer, la **Rocque** à la	land: N of Cobo (rue des Carterets), C. rocque q.v.
Mercel, le	fields: la Houguette, SPB. Extinct surname **Mersel**.
Merez, les	house, fields: N of rue Cevecq, SSV. Probably 'marais', spelt **Meres**, 1555, **Mares**, 1535.
Merluquer, le	field: les Bailleuls, SA. Extinct surname **Murluquey**.
Merottes, les	field: la Longue Rue, SSV. Spelt **Merottes**, 1695. Marottes q.v.
Merrienne(s), le(les)	houses, fields: le Bigard, F; rue Poudreuse, SM. Surname extinct 15th c. **Merrienne**. (not to be confused with Merrien, a Breton surname of the 19th c.).
Mes (**Mr**) **De Lisle**, de	see **Lisle**, **De**.
Meslin, de	land: les Mielles, VdelE; le Gain, SPB; les Galliennes, T. After **Guillome Meslin**, 1549. Also spelt **Melin**.

Meslin, rue de	road: unlocated, SPB.
Messe, la	field: E of les Islets, SPB. Probably 'masse', mound, being adjacent to former wind mill (moulin à vent). masse, moulin à vent q.v.
Messière(**s**), la(les)	land, all parishes. 'very small holding'. Messuage. Also spelt **Maisière**, **Maizière**, **Mezière**.
Messières au **Mière**, les	land: le Marais, C. Mière Le q.v.
Messières des **Roberts**, les	field: les Domaines, SSV. Robert q.v.
Messières des **Roulias**, les	field: le Chêne, F. Roulias q.v.
Messières Fallaize, la	land: les Courtes Fallaize, SM. Fallaize q.v.
Messières Garis, les	land: la Haye du Puits, C. Garis q.v.
Messières Hervy, les	land: Jerbourg, SM. Hervy q.v.
Messières Mateline les	land: la Hougue Anthan, SPB. Matheline q.v.
Messières Ricart, les	field: Jerbourg, SM. Ricart q.v.
Messières Thomas de Vic, les	land: le Mont Crevelt, SSP. Vic q.v.
Messurier(**s**), au(des)	houses, fields: le Foulon, la Vrangue, SPP; les Messuriers, le Gron, le Trépied, SSV; la Mare, le Grais, SPB; les Mouilpieds, SM; les Fauconnaires, SA. Extant surname **Le Messurier**.
Messurier, **Hougue** au	land: unlocated. SA.
Meules, rue des	road: St Jacques, SPP.
Meunier, le	fields: les Bas Courtils, SSV. the 'miller'. Mainnel q.v.
Meurtrière, la ruette	roadway: parallel to present Upland Road, SPP, led to le Cimetière des Frères. 'death lane'. Abolished early 19th c.
Mezerel, le	field: le Vallon, SM.

Mezerette, la — fields: les Monts, SSP; le Closel, CduV. Diminutive of **Messière**. Also spelt **Messerel**, **Maiserette**, **Meserette**.

Michel, de — fields: les Clos Landais, SSV; le Long Frie, SPB; ruette Rabey, SM. Forename (m) **Michel**.

Michel Agenor, d' — field: les Mares Pellées, CduV.

Michel Dry, la **Hure** — land: N of la Hougue Taive, T.

Michelle, de — fields: les Clos Landais, SSV; les Brehauts, rue St Pierre, SPB; la Planque, T. Forename (f) **Michelle**.

Michelle Guilbert, de — field: unlocated, CduV.

Michel Vaucourt, de — field: les Galliennes, SA.

Mielle(s), la(les) — houses, land, fields: la Banque, l'Islet, SSP; la Grève, CduV; les Goubeys, les Prinses, VdelE; la Margion, SSV. 'sand hills'. Also **Grandes Mielles**, **Petites Mielles**, (large, small).

Miellette, la — area: E of le Déhus, CduV. Diminutive of **Mielle**.

Miellette, la **Croûte** — land: unlocated, CduV.

Mieille, au — field: les Blicqs, SA. Spelling error for **au Mière**, subsequent to 1616.

Mière, au — fields: les Banques, SSP; le Hurel, SA. Surname extinct 18th c. (extant Jersey) **Le Mière**. Also feudal holding of land, Fief au **Mière**, SPB.

Mière, la **Croûte** au — field: la Grand Maison, VdelE. Derivation as above. croûte q.v.

Mière, la **Terre** au — land: les Rohais, SPP. Derivation as above. terre q.v.

Mière, les **Messières** au — field: (Guillome Le Mière), le Marais, C. Derivation as above.

Mière, rue au — road: les Ozouets, SPP. Derivation as above.

Milieu, du	fields, various, all parishes. 'middle' field.
Mill street	see **Pavé**, le **Haut**.
Millesent, la **Planque Guillemine**	field: E of les Massies, SSV. Surname extinct 15th c. planque q.v.
Millets, les	land: les Canichers, SPP. Extinct surname **Millais**.
Miquet, de	field: la Corbière, F. Forename (m) **Michel**.
Misère, la **Vallée** de	land: E of Queen's road, SPP. 'troublesome'. Recorded **Myssayre**, 1574.
Mizanne, la	land, fields: les Vardes, la Vrangue, SPP; le Mont de Val, C; les Heches, SSV. 'mezzanine' level.
Moigne, au	land: les Bergers, C. Extant surname **Le Moigne**. Also feudal holding of land, Fief **Dom Jean Le Moigne**, C.
Mollet, rue	road: W section of route du Braye, SSP. Now called route du Braye. Extant surname **Mollet**.
Monamy, la **Maison**	house: St Jacques, SPP. After **Andre Monamy**, 18th c. Surname extinct 19th c. **Monamy**.
Monnaie, le	district: le Monnaie, SA. Surname extinct 16th c. **Monnet**. Monnet, val q.v.
Monnel, le **Vau** de	land: Pleinmont T. Derivation as above.
Mon Plaisir, de	houses: St Jacques, SPP; les Landes, F. 'my pleasure'. Late 18th, early 19th c. villas.
Mont(s), le(s)	mount, promontory, all parishes.
Mont au **Nouvel**, le **Clos** du	land: Richmond, SSV.
Mont d'aval, le	district: S of la Haye du Puits, C. 'promontory slope'. 20th c. misnaming of 'le **Mont de Val**'.
Mont de Val, le	fields: la Haye du Puits, C. mount 'on side of valley', now Mont d'Aval, q.v.

Mont, le **Gros** — land: N of Fermains, SPP; les Sommeilleuses, F. 'great promontory'.

Mont le Val, **Camp** du — see **Mont le Val**.

Mont le Val, le — fields: rue de l'Arquet, SSV. Mount 'on top of' valley.

Mont(s), with proper names, are listed below. See also individual entries.

Mont Ardel, le	field: les Osmonds, SSP.	Ardel q.v.
Mont Arrivé, le	district: SPP.	Arrivé, q.v.
Mont Auban, le	field: les Blicqs, SA.	Auban q.v.
Mont au **No(rd)**, le	fields: Jerbourg, SM.	Nô q.v.
Mont au **Nouvel**, le	land: Fort Richmond, SSV.	Nouvel q.v.
Mont Blicq, le	fields: les Nicolles, SM.	Blicq q.v.
Mont Chinchon, le	land: le Catioroc, SSV.	Chinchon q.v.
Mont Crevel, le	land: Chouet, CduV.	Crevel q.v.
Mont Crevel(t), le	fort: South Side, SSP.	Crevel q.v.
Mont Cuet (**Cuhouel**) le	land: Chouet, CduV.	Cuhouel q.v.
Mont de la **Glie**, le	land: les Terres, SPP.	Glie q.v.
Mont de **Val**, le	fields: la Haye du Puits, C.	Val q.v.
Mont des **Moulins**, le	hill: les Grands Moulins, C.	Moulin q.v.
Mondril (**Mont Dry**), le	land: les Clercs, SPB.	Dry q.v.
Mont Durand, le	district: SPP.	Durand q.v.
Mont Durant, le	houses: Mont Durand, SM.	Durand q.v.
Mont Frie, le	house: Calais, SM.	Frie q.v.
Mont Gautier, le	cliffland: le Gouffre, F.	Gautier q.v.
Mont Gibel, le	hill: SPP.	Gibel q.v.
Mont Guilbert, le	fields: le Portelet, T.	Guilbert q.v.
Monts Héraults, les	fields: Monts Héraults, SPB.	Hérault q.v.

Mont Hubert, le	fields: la Fallaize, SM. Hubert q.v.
Monmains (**Mont Mains**), le	land: South Side, SSP. Mains q.v.
Mont Marché, le	district: S of airport, F.
Mont Morin, le	district: Delancey, SSP. Mourin q.v.
Mont Murel, le	house: Farras, F. Murel q.v.
Mont Plaisant, le	house: les Generottes, C. 'pleasant hill'. Villa built 19th c. alongside older house, la **Trappe Berger** q.v.
Mont Saint, le	district: le Mont Saint, SSV. Saint q.v.
Mont Sause, le	land: Icart, SM. Sause q.v.
Mont Tardif, le	fields: les Nicolles, SM. Tardif q.v.
Mont Val, le	house: le Bordage, SPB. late 19th c.
Mont Varouf, le	fields: rue de l'Arquet, SSV. Val q.v.
Mont Varouf, **Ruette** du	lane: near les Padins, SSV.
Monts, au **Pied** des	land: E of Delancey park, SSP. 'at the foot of the promontory'.
Monts, les **Cotes** aux	see **Aumont**.
Monterel, le	field: Jerbourg, SM. Diminutive of **mont**.
Montour, le	fields: les Houards, F. Origin unknown, spelt **Montour**, 1671.
Montville, de	house: les Vardes, SPP. Former 18th c. mansion burnt down late19th c.
Morin, le **Mont**	see Mont **Mourin**.
Morinel, le **Pont**	see **Mourinel**.
Morpaye, le	houses, fields: les Martins, SPB. 'dead land'. Recorded le **Mort Pays**, 1549.
Morts, rue des	road: to W of church, SA.
Morvels, le **Camps**	land: la Rochelle, CduV. Extinct surname **Mervel**. Recorded les **Camps Mervel**, 1591.

Motspertuis, le	see **Maupertuis**.
Motte, la	house: Saints, SM. 'mound, hillock'. Named in 18th c.
Mottes, la **Saudrée** ès	land: E of la Ramee, SPP.
Mouille, de	field: la Mare Pelley, les Trachieries, SSP. 'damp, wet'. Recorded **Moullier**, 1700.
Mouilpied(**s**), (les)	houses, fields: les Mouilpieds, SM; le Varclin, Bon Port, SM; les Hunguets, SA. Extant surname **De Mouilpied** (earlier **De Mouillepied**) 16th c.
Moulant, le **Clos**	land: Hacsé, CduV. Origin unknown. clos q.v.
Moulins à eau, (m) les	water mills. See also individual entries.

Site or ruin –

Moulin de la Mer	W of town church, SPP	Mer q.v.
Moulin du Milieu	Haut Pavé (Mill Street), SPP	Milieu q.v.
Moulin de Haut	la Charotterie, SPP	Haut q.v.
Moulin de Haut de la Vrangue	la Vrangue, SPP	Vrangue q.v.
Moulin de Bas de la Vrangue	la Vrangue, SPP	Vrangue q.v.
Moulin Foullerez	le Vauquièdor, SPP	Foullerez q.v.
Moulin de la Perelle	rue a l'Or, SSV	Perelle q.v.
Moulin de Beauvallet (**2**)	les Padins, SSV	Beauvalet q.v.
Moulin de Becquet	le Douit, SPB	Becquet q.v.
Moulins de Petit Bot (**2**)	Petit Bot, F/SM	Bot q.v.
Moulin Huet (**Luet**)	Moulin Huet, SM	Luet q.v.
Moulin de l'Echelle (**haut**)	les Galliennes, SA	Echelle q.v.

Building or substantial remains –

Grand (**Main**) **Moulin**	Grands Moulins (King's Mills), C	Grand q.v.

Moulin du Milieu	Grands Moulins (King's Mills), C	Milieu q.v.
Moulin de Haut	Grands Moulins (King's Mills), C	Haut q.v.
Moulin des Niaux	les Niaux, C	Niaux q.v.
Moulin de Quanteraine	le Bordage, SPB	Quanteraine q.v.
Moulin de l'Echelle (**bas**)	les Galliennes, SA	Echelle q.v.
Moulins à vent (m)	windmills. See also individual entries.	
Site only –		
Moulin de l'Hyvreuse	(Victoria Tower), SPP	Hyvreuse q.v.
Moulin des Vardes	les Vardes, SPP	Vardes q.v.
Moulin des Capelles	les Capelles, SSP	Masse q.v.
Moulin des Monts	Delancey park, SSP	
Mill	Baubigny, now a quarry, SSP	
Moulin de Noirmont	le Hamel, VdelE	
Moulin à vent	Castel School, C	Masse q.v.
Moulin des Grantez	rue de la Haye, C	Grantez q.v.
Mill	Rectory Hill, C	
Mill	site of KEVII hospital	
Moulin du Coudré	le Coudré, SPB	Coudré q.v.
Moulin à vent	les Islets, SSV	Islets q.v.
Moulin Domaine Dom Hue	(Airport), F	Hue q.v.
Moulin des Landes	les Landes, F	Landes q.v.
Mill	Mill lane, Rohais de haut SA	
Building or substantial remains –		
Moulin à vent	les Vardes, SPP	Vardes q.v.
Moulin à vent	Hougue du Moulin, CduV	Hougue q.v.

Moulin (cement mill)	les Hougues, C	Hougue q.v.
Moulin (cement mill)	le Mont Saint, SSV	Saint q.v.
Moulin De Sausmarez	Old Mill, les Camps du Moulin, SM	Sausmarez q.v.
Moulin du Hechet (Ozanne)	Steam Mill lane, SM	Hechet, Ozanne q.v.

Moulin(s), le(s) — houses, fields, all parishes. Adjacent to either water or wind mills.

Moulin à **vent**, **Camps** du — land: les Camps du Moulin, SM.

Moulin des **Niaux**, **Ruette** du — lane: Talbot Valley, C.

Moulin, **Douit** du — stream: Talbot Valley, C; near la Grande Rue, Bonvallet, Sous l'Eglise, SSV; le Douit, SPB.

Moulin Foulerée, **Hougue** du — land: unlocated, SA.

Moulin, **Hougue** du — land: Hougue du Moulin, CduV.

Moulin, la rue du — road: les Hougues Magues, SSP.

Moulins de la **Vrangue**, **Douit** des — stream: la Vrangue, SPP.

Moullin, d'**Abraham** — field: l'Echelle, SA. Derivation as above.

Moullin, de — field: la Vrangue, F. Derivation as above.

Moullin, de **Jean** — field: le Camp, C. Extant surname **Moullin**.

Moullin, de **Pierre** — fields: Choisi, SPP; le Camp Tréhard, la Contrée de l'Eglise, SA. Derivation as above.

Mount Durand — see **Durand**, **le Mont**.

Mount Hermon — see **Hermon**, **le Mont**.

Mount Row — road: W of le Mont Durand, SPP. Translation of 'mont'.

Mourain, Grand — land: W of Saumarez Park, C.

Mourain(**s**), les	house, fields: le Douit, VdelE; Saumarez, Sous l'Eglise, C. Surname extinct 17th c. **Mourain**.
Mouranderie, la	house, land: le Gains, T. Extant surname **Mourant**.
Mourants, les	house, fields: les Mourants, SM/SA. Derivation as above.
Mourières, la **Saudrée** des	land: les Rohais, SPP.
Mourin, de	fields: Airport, SSV/F/SA. Surname extinct 17th c. **Mourin**.
Mourin, **Hougue** de	land: les Godaines near Havelet, SPP.
Mourinel, le **Pont**	fields: la Haye du Puits, C. Extinct surname 15th c. **Mourinel**.
Moustier, la **Hougue** du	land: South Side, SSP. 'monastery'. Origin unknown, probably mis-named.
Mouton, la **Hure** au	land: la Caudré, SPB.
Moutonerie, la	house: les Rouvets, now la Parchonerie, VdelE. 19th c. nickname 'sheep'. Ouets, la Cote ès, q.v.
Moye(**s**), la(les)	1. district: E of l'Ancresse, CduV. Surname extinct 15th c. **De La Moye**. hougue q.v. 2. headland stacks: la Moye, F; Icart, Jerbourg, SM.
Mulettes, la rue des	road: now les Amballes, SPP. 'mule'. Possibly a difficult pack route or way.
Mur, le	1. embankment (now destroyed): W of le Camp du Roi, VdelE. 2. rampart: Le Mur (on which is Doyle column), Jerbourg.SM. Doyle, Jerbourg q.v.
Murel, le **Mont**	land: Farras, F. Surname extinct 17th c. **Mourel**. hougue q.v.
Murie, le pré de	land: le Douit, SPB. Origin unknown.
Murier, le	house, land: St Clair, SSP. Origin unknown, spelt **Meusrier** 18th c.

Musel, le **Blanc**	land: now Rockmount hotel, Cobo, C. 'white shrew'.
Mustel, de	land: les Talbots, C. Surname extinct 17th c. **Mutel**.
My Lady	see **Fee farms**, **Marais**, **le Rond**.

N

Naftiaux, les	district: les Naftiaux, SA. Extant surname **Naftel**, after **Collas Navetel**, 1607.
Nant, **Bordage Ricart**	feudal holding of land: la Rochelle, CduV. Extant surname **Nant**, spelt **Naont**, **Naoun**, **Naount** in 14th c.
Nathan, de	field: les Caches, SPB. Forename (m) **Nathan**.
Naunage, Fief du	feudal holding of land, C.
Naveaux, à	field: les Tilleuls, VdelE. Plural of **Navet**. 17th c. name.
Navet, de	fields: la Haye du Puits, C; la Fosse, Saints, SM. Surname extinct 15th c. **Navet**. rue q.v.
Navet, **Ruette** du	land, road: near Saints, SM.
Naye, de	see **Nez**.
Néel, d'**André**	fields: les Poidevins, SA. Surname extinct 16th c. **Néel**.
Néel, la **Croûte** de **Thomas**	field: les Camps. SM. As above. croûte, Niaux q.v.
Nermont, de	see **Noirmont**, de.
Nermont, la **Hougue** de	quarry: les Petils, CduV. Name taken from **De Nermont**, family name locally prominent in the 12th, 13th & 14th c. See Noirmont. hougue q.v.
Neuf(**m**), **neuve**(**f**)	'new', with proper name listed by parish. Se also individual entries.

SPP	Neuve Rue, Neuve Ville (New Town).
SSP	Neuve Rue.
CduV	Neuf Clos, Neuf Courtil.
VdelE	none.
C	Neuf Courtil, Neuve Maison.
SSV	Neuf Chemin, Neuf Clos, Neuf Courtil, Neuve Charrière, Neuve Maison.
SPB	Neuf Courtil, Neuve Maison.
T	Neuve Maison.
F	Neuve Courtil, Neuve Rue.
SM	Neuf Courtil.
SA	none.

New –
- New Place: Vauvert, SPP.
- New Road: South side, SSP; l'Aumone, C; le Manoir, F.
- New Street: rue Marguerite q.v., SPP.
- New Town: Neuve Ville q.v. Area centered on Allez, Havilland, Sausmarez and Union streets, SPP.

Nez, les — fields: la Fallaize, SM. Surname extinct 15th c. **Nes**, **Nez**. Also spelt **Naye**, **Nees**, **Nets**. See above **Néel**.

Nez, la **Croix** — field: S of la Ronde Cheminée, SSP. Derivation as above. croix q.v.

Nez, la **Croûte** de **Thomas** — field: les Camps, SM. Derivation as above. croûte q.v.

Nez, la **Hougue** — land: unlocated. SSP.

Nez, le **Bec** du — promontory: E of la Bouvée, SM. 'snout of a nose'.

Nez, les **Pièces** — land: les Vesquesses, C. Derivation as above.

Niaux, les — houses, fields, water mill: Talbot Valley, SA. Plural of extinct surname **Néel** q.v. Moulin à eau q.v.

Nicet, le — point of land: l'Eree, SPB; fields, Pleinmont, T. Valniquet. Surname extinct 16th c. **Nicel**.

Nicolas de **Jersey**, de — field: la Girouette, SSV. Jersey de, q.v.

Nicolas De Lisle, de	field: les Islets, SPB. Also noted as '**Collas De Lisle'**. Lisle de, q.v.
Nicolas, la **Pièce**	land: la Croix Bertrand, SM.
Nicolas, la **Pièce St**	land: la Hougue au Comte, C. Presumably a rente owed to a church confraternity.
Nicolle, **d'Anne**	land: Rocquaine, SPB. **Dan** (Dominus) **Nicollas** (surname not recorded).
Nicolle, de **Colle(ne)tte**	field: les Traudes, SM.
Nicolle, de **Jacques**	field: Baugy, Cdu V.
Nicolle, la **Croix d'Anne**	see **Revel**, **Dan** (Domninus) **Nicollas**. croix q.v.
Nicolle, la rue de **Dan**	road: now rue de l'Issue, SPB. Derivation as above.
Nicolle, les **Camps Collette**	land: W of Mon Plaisir, SPP. 'strips of land of **Collette Nicolle**'.
Nicolles, **Fontaine**	spring: unloc. SM.
Nicolles, **Hougue** des	land: N-E of les Osmonds, SSP.
Nicolles, la **Croix** des	site, house, land: les Beaucamps, C. Former wayside cross. Forename (m) **Nicolas**. croix q.v.
Nicolles, la **Croûte** des	land: W of la Ville au Roi, SPP; les Nicolles, SSP. Derivation as above. croûte q.v.
Nicolles, les	houses, land: Fort George, SPP; les Osmonds, l'Islet, les Gigands, les Nicolles, SSP; les Delisles, les Grands Moulins, C; la Hougue Fouque, SSV; les Nicolles (les Nicollets), F; les Poidevins, SA. Extant surname **Nicolle**.
Nicollets, les	house, land: les Nicolles, F. Recorded les Nicollets, early 19th c.
Nid de **l'Herbe**, le	land: l'Ancresse, CduV. 'niches of grass'. Recorded as **Nicqs de l'Herbe**, 1591.
Niots, les	see **Niaux**.

Niquet, le **Val**	fields: Pleinmont, T. Surname extinct 16th c. **Nicel**.
Nô, au	fields, land: Côbo, C; les Rouvets, SSV; Jerbourg, SM. Extinct surname **Nô**, **Nos**. Also mis-spelt **Nord**, 18th & 19th c.
Nô, le **Val** au	land: S of le Bouet, SPP. Derivation as above.
Nocq, le	sluice: rue du Nocq (Nocq road), SSP. Sluice fitted at corner of St Sampson's bridge to control flow of water in former canal along N side of Nocq road and la Grosse Hougue, connecting the Salt Pans with the Harbour, 1806–1840s.
Noel, de **Henry**	field: la Vieille Rue, SSV. Surname extinct 21stc. **Noel**.
Noel, de **Jean**	fields: la Fontaine, les Prevosts, les Jetteries, le Gron, SSV. Derivation as above.
Noel, de **Louis**	field: Airport, SA. Derivation as above.
Noirmont, de	fields: les Hougues, les Vardes, SSP. Surname extinct 17th c. **De Nermont**. Spelt **De Noirmont** since 19th c. Also feudal holding of land, Fief de **Richard De Nermont**, VdelE.
Noirmont, de	house, land: Pleinheume, VdelE. Derivation as above.
Noirmont, **Hougue** de	land: E of le Dehus, near the beach SSP.
Noirmont, le **Vingaine** de	name given to detached area of parish of St Sampson, currently disused. Name taken from **De Nermont**, family name locally prominent in the 12th, 13th & 14th c. Feudal holding of land in the same area, Fief des **Bruniaux de Noirmont** (de Nermont) SSP. Vingtaine q.v.
Nord, au	see **Nô**.
Nord-Est, le **Bordage**	feudal holding of land, le Dos d 'Ane, C. Bordage q.v.

Nordest, la **Vallée**	land: les Vesquesses, C.
Norgeot, la **Terre**	houses, fields: la Terre Norgeot, SSV. Derivation as above.
Norgiots, les	house, district: les Norgiots, les Eperons, SA. Surname extinct 16th c. **Norgeot**, also spelt **Norgiot**, **Norjiot**, **Norgot**.
Norman, la **Croûte**	field: les Pièces, F. Extant surname **Norman**.
Norman, la **Pièce**	land: la Rivière, C. Derivation as above.
Normanville, de	house, land: la Couture, SPP. Surname extinct 16th c. **De Normantville**. Ref. **Lorenche de Normantville** (contract) 1485.
Nort, **Camp** au	see **Nô**.
North Esplanade	thoroughfare and gardens: opposite Careening Hard SPP. Built 1850s.
Notre Dame, de	fields: Candie, C; les Laurens, T; les Truchots, SA. 'Our Lady'. Presumably rentes owed to church.
Notre Dame de l'**Epine**	see **Epine**, la **Chapelle** de.
Notre Dame de la **Perelle**	see **Ste Appoline**.
Notre Dame du **Mont Gibel**, **Fontaine**	spring: at top of Arcade Steps. SPP.
Nouel, de	see **Noel**.
Nouettes, les	fields, land: N of les Landes, F. 'marshy land'. Spelt **Nouettes**, 1504. fontaine q.v.
Nouettes, **Fontaine**	spring: W of la rue Perrot, F.
Nouis, les	fields: la Hougue Patris, CduV; Beauvalet, SSV. Origin unknown.
Nouis, la **Vallée** des	land: S of the reservoir, SSV.
Nouvel, le **Mont**	land: Fort Richmond, SSV. Extinct surname **Le Nouvel**.
Novelle, la rue à la	land: les Forfaitures, C. Derivation as above.

O

Oberlands,	house, land: W of les Frietiaux, SPP. 19th c. villa.
Obits, les	fields: la Hougue Fouque, SSV; Airport, les Rohais, SA. 'obits'. i.e. from rentes payable for requiems. Also spelt **Lobi, Loby, Obis**.
Oie, à l'	meadow: le Groignet, C. 'goose'. A wet area.
Oignonière, l'	fields: le Jamblin, CduV. 'onion patch'. Spelt l'**onnière**,1654.
Oisel, l'	field: S of la Grande Mare, C. Surname extinct 14th c. **Loisel**.
Olimpe, d'	former field: South Side, SSP. Forename (f) **Olimpe**.
Ollivier, d'	fields: les Annevilles, SSP; les Niaux, C. Forename (m), or extant surname **Ollivier**.
Ollivier, de **Collas**	field: les Varendes, C.
Ollivier, la **Croûte** d'	field: les Nicolles, F. Derivation as above.
One, l'	see **Allonné**.
Or, à l'	field: les Barrats, VdelE. 'golden' (brown gravel).
Or, la **rue** à l'	road: W of King's road, SPP; S of le Mont Saint, SSV. Derivation as above.
Or, les **Rocques** à l'	field: N of rue à l'Or, SPP; Pleinmont, T. Derivation as above.
Or (le **Vauquiéd**-)	valley: Havilland Hall, SPP. 'valley which is of gold'. Earliest ref. 1504 'golden (brown gravel)'. val q.v.
Orais, les	former fields: N of les Cherfs, C; les Fauconnaires, SA. Surname extinct 15th c. **Ysorey**. Also spelt Yzorey. Ysorey, marais q.v.

Orais, route des	road: formerly rue Ysorey, now route de Côbo.
Orange Belle, d'	land: les Monts Héraults, SPB. Forename, probably corrupted. Recorded **Orrenge Belle**, 1661. Belle q.v.
Orangers, les	field: la Rochelle, CduV. Surname extinct 15th c. **Errengiers**. Recorded **Errengiers**, 1654, 1755. **Orengiers**,1809.
Orete, la **Chapelle** de l'	see **Lorette**.
Oreys, les	see **Orais**.
Orge, à l'	fields: Baubigny, SSP; les Landes, C. 'barley'.
Orgeris, **Douit** des	stream: from the Manor of Anneville to the Braye du Valle, SSP. Tonelle q.v.
Orgeris, la rue des	road: Fermains, SPP.
Orgeris, les	fields: les Grandmaisons, SSP; S of les Landes, CduV; le Hamel, VdelE; W of les Adams, SPB; E of route de Sausmarez, SM. Area of 'barley'.
Orgeuil, le **Château** d'	castle: 'pride castle', alternative name for **Château** des **Marais**, SPP. château q.v.
Orgeuil, le **Marais** d'	marsh: alternative name for the area adjacent to above Château. 13th & 14th c. name for both above. marais q.v.
Ori, le **Camp**	land: les Hautgards, CduV. Probably from extinct surname. 20th c. ref. only.
Ormes, les	land: les Effards, SSP; la Neuve Maison, SPB; les Rohais, l'Echelle, SA. 'elms'.
Orvillière, l'	house, field: les Bruliaux, SPB. Origin unknown. Recorded l'**Oreulièrre**, 1685, l'**Orvellière**, 1754.
Osiers, les	field: les Closios, CduV. 'willow'.
Osmonds, la **Hougue** des	houses, land: S of les Marettes, SSP. Extant surname **Aumont**.

Osmonds, les	fields: E of les Marettes, SSP. Extant surname **Aumont**. Also spelt prior to 18th c. les **Aumonts**. Aumont q.v.
Oso, le **Camp**	fields: E of le Déhus, CduV; Richmond, SSV. 'bony'. (osseux, ossu) i.e. hard.
Ossillons, les	land: coast, les Bas Marais, SSV. Diminutive of **Oso** above. 'bony furrows'.
Ouar, d'	see **Houards**.
Ouets, les **Côtes** ès	house: les Rouvets, VdelE. 19th c. nickname 'geese'. Moutonnerie, q.v.
Ovine, d'	field: les Baissieres, SA. Origin unknown. Spelt **Odovere**, 1701. **Ovière**, 1793,
Ozanne, d'	fields: les Grantez, les Roussiaux, Woodlands, C; les Bourgs, SA. Extant surname **Ozanne**. Spelt **Ozenne** prior to 18th c. as in Normandy.
Ozanne Mill	wind mill: 19th c. moulin à vent, also called le **Hechet** mill q.v.
Ozon, le **Camp**	see **Oso**.
Ozouets, les	house, fields: la Couture, SPP. Owned by **Nicollas Osouet** junior, 1574. Surname extinct 17th c. **Osouet**.

P

Paas, la rue des	green lane: S of Rocquaine, towards les Cambrees, SPB. 'path, passage.'
Paas, les	houses, fields: le Marinel, SPB.
Paas, ruette des	see **Paas**, la **rue** des.
Padart, les **Camps**	land: E of la Couture, SPP. Surname extinct 16th c. **Padart**. camps q.v.
Padins, les	houses, land: le Mont Varouf, SSV. Possibly 'small walk'. Earliest reference 1822.

Padis, les — land: l'Hyvreuse, Vauvert, Havelet SPP. Origin unknown.

Page(s), les — houses, fields: Fort George, SPP; le Préel, C; les Sages, T; Les Pages, SM. Extant surname, **Le Page**.

Page, la rue au — road: la Hougue Fouque, SSV. Derivation as above. rue q.v.

Pageot(s), les — land: Brock road, SPP; les Grandmaisons, SSP. Surname extinct 17th c. **Pageot**.

Pains, les **Clos** — district: N of school, CduV. Extant surname **Paint**. clos q.v.

Paint, la **Fosse** — land: les Blanches, SM. Derivation as above. fosse q.v.

Paints, la **Rocque** aux — land: la Soucique, F. Derivation as above. rocque q.v.

Paint, le **Vau** — field: le Vaupaint, rue à Gots, C. Derivation as above. val q.v.

Paintain, la rue de — see **Panthon**.

Paisan(t), la **Croix** — see **Paysans**.

Paisant, la **Croix** — site: on crossroads E of la Fontaine, SSV. Former wayside cross. Surname extinct 15th c. **Du Paysant** (**Du Paisant**). croix q.v.

Paison, le **Courtil** — former field: les Girards, C. Recorded **Courtil Paison**, 1624, **Courtil Poisson**, 19th & 20th centuries.

Paix, le **Courtil** à — house, land: les Annevilles, SSP. 'peace'. Presumably reference to rente in memory of deceased. Also spelt **Pois**, **Poix**.

Palerons, les — fields: Fort George, SPP; rue de l'Arquet, SSV. 'shoulder' of land.

Pal(l)ière, la — fields: N of Cocagne, CduV; W of les Nouettes, F. 'upper level, top part'.

Pal(l)ot, de — fields: les Buttes, la Palloterie, SPB. Extant surname **Palot** (Jersey spelling Pallot).

Pal(l)oterie, la	district: W of les Bruliaux, SPB. Derivation as above.
Palotte, la	meadow: les Landes, C. 'off colour' vegetation.
Pandant, le	see **Pendent**.
Panier, le **Vau**	land: NE of Queen's road, SPP. Surname extinct 15th c. **Pasnier**.
Pannelottes, les	fields: les Bas Courtils, SSV. 'small reclaimed areas'.
Pans de terre (m)	land: late medieval 'pans' or plots of drained land, les Longs Camps (la Pitronnerie), SPP.
Panton, de	field: S of le Moulin de haut, C.
Panthon, la rue	road: green lane S of la Maison de haut to les Fauquets, C. Surname extinct 16th c. **Panthon**.
Pantlaron, le **Bordage**	Bordage: le Vauquièdor, SPP. Surname extinct in medieval period. Bordage q.v.
Paradis, de	houses, fields: les Vardes, Doyle road, SPP; N of la Rochelle, CduV. 'paradise', blessed land, garden, i.e.desirable.
Paradis, le **Hommet**	tidal islet: les Petils, CduV. Derivation as above. hommel, hommet q.v.
Parbougourd, de	field: la Passée, SSP. Origin unknown, first reference 1851.
Parc, le(s)	land, fields: la Turquie, CduV; les Goubeys, VdelE; slopes to E of Moulin de haut, C; Airport, F. 'pound', enclosure for pigs.
Parc au **Marchant**, le	land: la Hure Mare, CduV. Marchant q.v.
Parc Canu, le	land: la Croûte Becrel, CduV. Canu q.v.
Parc de l'**Epine**, le	field: Paradis, CduV. epine q.v.
Parc de la **Mielle**, le	land: la Rochelle, CduV. mielle q.v.
Parc de **Putron**, le	land: les Terres, SPP. Putron q.v.

Parc Meslin, le	land: les Mielles, CdelE. Meslin q.v.
Parchonnerie(s), les	houses, land: St Jacques, la Ramée, SPP; l'Islet, SSP; le Grand Marais, CduV; S of le Rocher, VdelE; les Grantez, C; les Camps, Sausmarez manor, Jerbourg, SM. Re 'parcener' (co-heir), implying ownership arising from inheritance. Also spelt **Parçonnerie**, **Parçonniers**, **Personnerie**.
Parcq, la rue du	street: la Charroterie, SPP. 'Park street'. Origin presumably from parc q.v.
Parcq, **Ruette** du	land: rue du Pré, SPP.
Parcs Capelles, les	land: la Moye, CduV. Owned by **Georges Capelle**, 1591. Capelle q.v.
Paré, de	fields: le Hurel, la Grande Rue, CduV. Surname extinct 17th c. **Parrey**.
Paré, la **Hougue**	field: la Moye, CduV. hougue q.v. Derivation as above
Paré, de **Leonard**	field: Hacsé, CduV. Owned by heirs of **Leonard Parrey**, 1591. Derivation as above.
Paré, le **Clos**	field: le Marais, C. clos q.v. Derivation as above.
Paresse, la	field: les Piques, SSV. 'idleness', i.e. neglected.
Paris, de	houses, fields, road: Paris street SPP. Surname extinct 15th c. **Paris**.
Paris, la **Croûte**	houses, land: N of Hougue du Valle, CduV. Derivation as above.
Parisienne, la	field: SA. Origin unknown, probably extinct surname.
Parquet, le	small area of land, all parishes. Diminutive of parc, small pound or enclosure for pigs.
Pas, le	see **Paas**, les.
Passée, la	house, land: La Passée, SSP. 'the beyond', i.e. at the extremity of the parish.

Passeur, au	road, field: rue du Passeur, CduV; les Caches, SM. Surname extinct 15th c. **Le Passeur**.
Passeur, **Camp** au	land: l'Ancresse, CduV; les Caches, SM.
Paté, le **Camp** du	field: les Issues, SSV; la Bouvée, SM. Surname extinct 15th c. **Le Pastey**. Also feudal holding of land, Bouvée **Henry le Pastey**, SM.
Patée, la	field: la Bouvée, SM.
Patourel, la	land: le Camp Tréhard, SA. Extant surname **Le Patourel**.
Patris, la **Hougue**	land: E of l'Ancresse, CduV.
Patris, la **Ville**	land: rue des Landes, SM. Owned by **Pierre Patris**, 1517. Surname extinct 16th c. **Patris**. ville q.v.
Patron, de	field: la Girouette, SSV. Surname extinct 16th c. **Patron**.
Paturage, le	fields: les Vallettes, SSV; les Corbinez, SPB; le Coemil, T. 'purprestre', i.e. illegal encroachment. Also spelt **Patures**.
Paul, de	see **Potellerie**.
Pau(l)mier, du	land, field: les Barrats, VdelE; la Palloterie, SPB. Surname extinct 17th c. **Le Paulmier**.
Pauvres, la **Maison** des	see **Maison des Pauvres**.
Pavé, le	house, land: la Porte, C. 'paved' way.
Pavé, le **Haut**	street, house, land: now Mill street, SPP; la Lande, C. 'high' pavement. Name changed early 20th c. rue q.v.
Pays, au **Rond**	land: N of Fermains, SPP. 'rounded' land.
Pays, le **Mort**	see **Morpaye**, le.
Paysans, les	district: les Paysans, SPB. Surname extinct 15th c. **Du Paysant (Du Paisant)**.
Payson, la **Rocque**	area: S of la rue Rocheuse, SPB. Recorded la **Rocque Poisson**, 19th & 20th centuries

	[printer's error, 1835]. Derivation as above.
Paysson, le **Clos**	field: les Sages, SPB. Recorded **Clos Poisson**, 19th & 20th centuries. Derivation as above.
Pedvin street	see **Poidevin**, la **rue**.
Pelichon, **Fontaine**	spring: unlocated, VdelE.
Pellerin, de	field: rue du Val, SPB. Surname extinct 14th c. **Pelryn**.
Pelley, le	fields: Fort George, SPP; Bordeaux, CduV; les Pelleys, la Hougue du Pommier, C.; Pleinmont, T; les Eperons, SA. Extant surname **Le Pelley**. Also spelt au **Pellé**, la **Pellée**. Also feudal holding of land, Fief **ès Pellais**, C.
Pelley, **Clos** au	land: le Hougue du Pommier, C.
Pelley, **Camp** du	land: top of Pleinmont, T.
Pelley, **Douit** du	field: near the coast at Cobo, C.
Pelleyrie, la	house: la Contrée de l'Eglise, SA.
Peltier, de	fields: E of les Capelles, SSP; le Prevôté, SPB; field, les Adams, SPB. Late of **Pierre Le Pelletier** fils Collas, 1536. Surname extinct 16th c. **Le Pelletier**.
Peltier, **Rogier Le**	field: le Lorier, SPB (early 16th c. ownership). Derivation as above.
Pénage, le	fields: les Caches, SM. 'wooded scrub' for pigs.
Pendant(s), le(s)	land, fields, all parishes. Steep rocky slope. Also spelt **Penchant**.
Pendant des **Puits**, le	field: Castel Hospital, C. puits q.v.
Pendant des **Vallées**, les	field: les Eturs, C. vallees q.v.
Pendant de la **Varde**, la	land: Pleinmont, T. varde q.v.
Pendant Heaume, le	field: les Villets, F. Heaume q.v.

Pendante, la **Rocque**	see **Rocque**.
Penel, le	see **Pesneille**.
Penelle, la	see **Pesneille**.
Pépinière, la	land: numerous, SPP; C; SM; SA. 'nursery garden'.
Pequerie(s), la(les)	land: S of la Passée, SSP; la Rousse Mare, SPB. 'fishery'. fish drying area.
Perchard, de	way, field: Perchard's lane, SPP; les Eturs, C. Extant surname **Perchard**.
Perchie, la	fields: Castel school, les Fries, le Marais, C. 'perches, racks' to dry crops. Misspelt **Percée**.
Père, de	see **Perre**.
Père Le Metre	field: les Rocques, T, from Pierres **Le Maitre** q.v.
Perelle, la	1. district, bay: SSV. 2. Chapelle de Notre Dame de la Perelle, see Ste Apolline. 3. Moulin à eau de la Perelle, see Moulin. 4. coastal battery: mid-bay, 18th c. Recorded de Perrella, 1157. Origin unknown.
Pères Bourgois	land: les Nouettes, F, from Pierres **Bourgeois** q.v.
Pères Le Fevre	field: les Grantez, C, from Pierres **Le Feyvre** q.v.
Perière, de	field: le Russel, F. Origin unknown.
Perin(ne), de	see **Perrin**.
Peringier, le	field: la Longue rue, SSV. Origin unknown
Perotin, le	see **Perrotin**.
Perques, le **Courtil** des	field: les Camps, SM. After **Pierre Des Perques**. Extant surname **Desperques**.
Perquesse, à la	see **Vesquesse**.

Perquin, de	land, fields: King's road, St Jacques, l'Hyvreuse, SPP; les Fourques, SA. Sale of rente of late **Thomas Perrequin** 1563. Surname extinct 16th c. **Perquin**. Above surname may be a gallicized form of 'Perkins'.
Perre, à la	houses, land, fields: la Hougue à la Perre, SPP; les Hougues Peres, CduV; le Préel, C; les Rebouquets, SM; le Tertre, C.
Perre, Guillaume La	fields: la Maison de Haut, SSV. Late of **Guillaume La Perre**, 1551. Surname extinct 18th c. **La Perre**. hougue q.v.
Perre, Hougue à la	see **Perre**.
Perres, Hougues	see **Perre**.
Perrin, la **Croûte**	field: les Trachieries, SSP.
Perrin, le **Pont**	land: N of les Rouvets, VdelE. Surname, **Perrin**. croûte, pont q.v.
Perrinne, de	field: la Fosse, SM. Origin unknown.
Perrinne, la **Croix**	site: la Pointe, SPB. Former wayside cross. croix q.v.
Perrins, la **Croûte**	see **Perrin**, la **Croûte**.
Perron, le	fields, land: N of le Mont Durand, SPP; la Mare de Carteret, C; les Sages, les Simons, les Laurens, T. 'small standing stone, boulder'.
Perron de **Laitte**, le	land: rue de Laitte, SPB. Derivation as above.
Perron du **Roi**, le	stone: le Bourg, F. On le chemin le Roi, and formerly connected with la Chevauchée St Michel. Derivation as above.
Perrons, les **Vaux**	field: les Etibots, SPP. Derivation as above. val q.v.
Perrot, la rue	road: Farras, F. rue q.v.
Perrot Heaume, le	field: les Houards, F. Forename (m) **Perrot**, diminutive of **Pierre**.

Perrotte, de — fields: le Groignet, la Riviere, C. Forename (f) **Perrotte**, from diminutive of **Pierre**.

Perrottes, **Fontaine** — spring: S of les Landes, CduV.

Perruque, la — former field: W of Castel hospital, C. 'young lawyer', as a nickname.

Personnerie, la — see **Parchonnerie**.

Personnerie, la rue de la — track: les Camps, SM. see **Parchonnerie**.

Pesneille, de — fields: le Mont Morin, SSP; les Fauquets, C; les Eperons, SA. 'pile of tatters', i.e. crop remnants. Also spelt **Penel**, **Pesnelle**.

Petils, les — land, fields: N of Cocagne, CduV; S of les Jetteries, SSV; Pleinmont, T. 'compacted' land.

Petiot, de — land: la Couture, N of l'Hyvreuse, SPP.

Petiot, la **Croûte** — land: King's road, SPP. Surname extinct 16th c. **Petiot**. croûte q.v.

Petit, label attached to substantive name to indicate a small field of that name.

Petit Bot, de —
1. land: 2 valleys, S of le Bourg, F; S of les Martins, SM.
2. 2 former water mills, moulins à eau: Petit Bot, upper and lower at foot of valleys. 'small wood', the lower valleys wooded.

Petit, la **Ville** au — see **ville**.

Petit, le **Rouge** — field: les Fries, C. probably 'compacted' land. Recorded **Rouge Paetis**, 1470.

Petit Marais, le — see **Marais**, le **Petit**.

Petit Port, le — see **Port**, le **Petit**.

Petite Porte, la — see **Porte**, la **Petite** (under **Port**, le **Petit**).

Petites Portes, les — see **Portes**, les **Petites**.

Petris, les **Rouges** — field: S of le Long Trac, SM. 'reddish, compacted soil'. Recorded de **Rogez Petilz**, 1488. petils q.v.

Pettevin(s), les — see **Poidevin(s)**.

Pezeril, **Croûte** du — land: le Pezeril, CduV.

Pezeril(s), le(s) — land: Chouet, CduV; Pleinmont, T. 'perches, racks' for drying fish. Eperquerie q.v. Also **Fort des Pezerils**, 18th c. battery.

Pezette, la — fields: la Ramee, SPP. Extant surname **Pezet**.

Philemon, de — field: le Malassis (la Brigade), SA. After **Philemon Effart**, early 17th c.

Philemon, la **Croûte** — field: N of les Nicolles, F. Forename (m) **Philemon**.

Philippe, de — fields: the Track, SSP; les Padins, SSV; rue Cauchez, SM. Surname extinct 15th c. **Philippe**. Also feudal holding of land, Fief des **Philippes** (late of Pierre Philippe), SSP.

Philippe, le **Douit** du **Vau Dan** — see **Dan**, **douit**, **val**.

Philippe Nicolle, de — field: les Coutanchez, SPP.

Philippin, de — field: Icart, SM. Forename (m) **Philippin**.

Piart, la **Croûte** — land: Fort George, SPP. Surname extinct 14th c. **Piart**.

Piart, le **Marequet** — land: la Rochelle, CduV. Derivation as above.

Picart, au — field: les Poidevins, SA. Surname extinct 14th c. **Picart**.

Picart, la rue — road: la Villette, SM. Derivation as above.

Pichet, le **Clos** — see **Picquet**.

Picot, de — land: le Truchot, SPP. Extant surname **Picot**.

Picquerel, le — land: on coast, le Houmet, VdelE. 'point of land'.

Picques, les	house: N of le Gron, SSV. Origin unknown, name first recorded 1797.
Picquet house	building: by Town church, SPP. "Erected by Government AD 1819, M-Genl. Bayly Lt. Govr." Used for picket duty by the British garrison.
Picquet, le **Clos**	field: Jerbourg, SM. Surname extinct 20th c. **Picquet**.
Pie, la **Croûte** à la	land: Belval, CduV.
Pies, ès	houses, land, fields: Belval, CduV.
Pies, la **Ville** ès	see **ville**. Surname extinct 17th c. **La Pie**.
Pièce(s), la(les)	houses, land, fields, all parishes. Unenclosed 'piece' of land.
Pièce, la **Carrée**	fields: les Houmets, C; le Mont le Val, rue de l'Arquet, SSV. 'square' piece of land.
Pièce, la **Grande**	fields: les Vallettes, SSV; les Laurens, T; le Coin Colin, SM. 'large' piece of land.
Pièce, la **Longue**	fields: la Haye, C; les Lohiers, rue de l'Arquet, SSV; Rougeval, T; les Bailleuls, les Baissieres, SA. 'long' piece of land.
Pièce, la **Platte**	field: les Houmets, C. 'flat' piece of land.
Pièce, la **Ronde**	field: N of la Traude, SM. 'rounded' piece of land.
Pièce, la **Torte**	fields: les Pelleys, la Hougue, C. 'twisted' piece of land.

Pièce(s) with significant proper names. See also individual entries.

Pièce de Banque	SPB	**Pièces Nez**	C
Pièce des Bergerots	C	**Pièce Nicolas**	SM
Pièce de Birro	VdelE	**Pièce Norman**	C
Pièce des Cheliziers	SPB	**Pièce St Jacques**	T
Pièce à Chiens	C	**Pièce St Nicolas**	C
Pièce ès Cames	C	**Pièce de Trésor**	SPB

Pièce Gruchy	C	**Pièce du Val**	C
Pièce des Ligneris	C	**Pièce Vallen**	C
Pièce de la Maison Dugnet	C		

Pièces, les **Courtes** fields: la Haize, la Lande, CduV; les Pelleys, C. 'short' piece of land.

Pied de l'**Epine**, le land: Pleinmont, T. 'at foot of' the thorn.

Pied des **Monts**, le land: les Grandmaisons, SSP. 'at foot of' the mount.

Pied du **Mur**, le land: Jerbourg, SM. 'at foot of' the rampart.

Pierre, de land: la Longue Pierre (les Cherfs), C.

Pierre, de field: rue de l'Arquet, SSV. 'standing stone', megalith. chaise, perron, rocque q.v.

Pierre au **Cucu**, la land: les Huriaux, former l'Erée aerodrome, SPB. Cucu, le q.v.

Pierre Belle, le **Bordage** feudal holding of land, W of King's Mills, C. Bordage q.v.

Pierre Cocquerel see **Cocquerel**.

Pierre, **Dan** (d'Ampiere) see **Dan Pierre**.

Pierre De Baugy see **De Baugy**. **[Baugy, De]**

Pierre de l'**Hyvreuse**, see **Hyvreuse**, l'.

Pierre Du Val see **Val, Du**.

Pierre Fallaize see **Fallaize**.

Pierre Guillard see **Guillard**.

Pierre Henri see **Henry**.

Pierre, la **Grosse** stone: formerly at la Hougue à la Perre, SPP. 'large upright boulder'.

Pierre, la **Longue** land: S of les Fries, Retot, les Grantez, les Cherfs, St George, C; (l'Essart) les Marettes, les Fontenelles, les Massies, SSV; W of le Felconte, les Paysans, SPB; N of la Croix au Baillif, SA. 'tall standing stones'. rocque, la longue q.v.

Pierre Le Patourel see **Patourel, Le**.

Pierre Martin see **Martin**.

Pierre Percée, la house: le Mont Durand, SPP. A 'pierced' standing stone.

Pierre Robilliard see **Robilliard**.

Pierre Rouget see **Rouget**.

Pierre Du Val see **Val, Du**.

Pierres, les **Blanches** land: S of les Hubits, SM. 'white (quartz) stones'. Possibly a former megalith.

Piette, la houses, land: W of la Salerie, SPP. Surname extinct 16th c. **Piette**.

Piette, la **rue** road, field: S of Castel church, C/SA. Derivation as above.

Pignon, le houses, field: les Coutanchez, SPP; les Arguilliers, VdelE; les Pelleys, la Pouquelah, les Landelles, le Camp, C; le Coudré, SPB; les Sages, T. 'gable' of a house, building.

Pigny, de field: la Turquie, CduV. Surname extinct 16th c. **Pignel**.

Pigny, la **Croûte** field: la Turquie, CduV. Derivation as above.

Pilbot, le fields: les Quartiers, les Osmonds, SSP. Surname extinct 15th c. **Pillebot**.

Pillory, la **Place** du site: 'la grande Rue, proche la meyr' (Quay street), SPP. Reference 1503. Cage or stocks for malefactors.

Pinant, la **Hougue** field: les Beaucettes, CduV. Origin unknown. Earliest record 1755. hougue q.v.

Pins, les house, field: E of Côbo, C. 'pine'. Recorded 1551.

Pipet, de fields: les Videclins, C; les Norgiots, SA. Extant surname **Pipet**.

Pipet, les **Vallées** fields: Candie, C. Derivation as above. vallées q.v.

Piquenot, la **Fosse** — field: l'Alleur, la Hougue Anthan,T. Surname extinct 15th c. **Piquenot**.

Piquet, **Clos** du — land: unlocated. SM.

Pissot(s), le(s) — land: la Vrangue, les Terres, SPP; les Grands Moulins, les Hurettes, les Grantez, C. 'shallow well'. Pizode q.v.

Pistolet, le — field: les Loriers, CduV. 'odd shape'.

Pitard, de — fields: la Mazotte, CduV; les Mares, SPB; le Variouf, F. Surname extinct 17th c. **Pitard**.

Pitochon, de — field: les Arguilliers, VdelE. Origin unknown.

Piton, le **Clos** — field: les Forfaitures, C. Surname extinct 17th c. **Pithon**.

Pitouillette, **Douit** de la — field: les Hautes Vallées, C.

Pitouillette, la — house: field. Les Petites Vallées, C. 'marshy' land.

Pitronnerie, la — area: N of la Vrangue, SPP. 19th c. name (la **Putronnerie**) from **De Putron** q.v.

Pitrot, de — field: les Arguilliers, VdelE. Origin unknown.

Pizodes, les — fields: les Landes, rue Cohu, C. 'shallow wells'. Also spelt **Pizottes**. Pissot q.v.

Plaiderie, la —

1. site: old **Royal Court** building, off le Pollet, SPP.
2. field: W of les Hougues, SSP. Court place of Fief de **Henry De Vaugrat**.

Plaidy, le **Frie** — plot: les Pelleys, C. Court place of Fief **De La Cour**.

Plain(s), le(s) — land, fields: les Arguilliers, VdelE; Pleinmont, T; Jerbourg,SM. 'flat, open area'.

Plaisance, de — houses: les Videclins, C; former house, les Hechès, SPB. demolished in 1950s. 'elegant' house.

Planel, le	fields: W of le Grée T; unlocated, SM. Diminutive of plain, 'small flat, open area'.
Planitrel, le	land, field: la Maison de haut, SSV; Carteret, C. small areas of 'flat' land. Also plural, **Planitriaux**. **Platriaux** q.v.
Planque, la	houses, fields: les Baissières, la Ramée, les Rohais, SPP; le Douit, VdelE; les Grands Moulins, la Haye du Puits, C; le Mont Varouf, SSV; le Douit, SPB; les Rocques, T; Airport, F; la Quevillette, SM. 'plank' for crossing a stream (douit). pont q.v.
Planque Cabart, la,	see **Cabart**.
Planque de la **Longue Giolle**	see **Giolle**.
Planque des **Longs Camps**	see **Camps**.
Planque des **Marais**	see **Marais**.
Planque des **Rohais**	see **Rohais**.
Planque, **Douit** de la	field: S of les Ozouets, SPP.
Planque Guillemine Millesent	see **Millesent**.
Planque Hue,	see **Hue**.
Planquette, la	field: la Rivière, C. Diminutive of planque, 'small bridging plank'.
Plaquière, la	house, field: les Marais, SSP; les Genâts, C. 'flat spot'. Also spelt **Platière**.
Plantation, la	fields: la Ville au Roi, SPP; les Rohais, SA. 18th c. tree planting, in plantation.
Plat(te)s,	'flat', with significant proper name. See also individual entries.

Plats Camps, SPB; SM	**Platte Hougue**, CduV.
Plat Courtil, SPB; SA	**Plats Marais**, SSP
Platte Hache, SPP	**Plat Pré**, SSP

Plat Hommel, SSP — **Platte Rocque**, SM

Plat Houmet, SSP

Plataine(s), les — land: les Marais, SSP; la Haye du Puits, C; le Felcomte, SPB. small 'flat' area, enclosed.

Platel, le **Clos** — field: N of rue des Marais, VdelE. Surname extinct 15th c. **Platel**.

Platon, le — land: S of Clifton, SPP; W of St Clair, la Ronde Cheminée, Baubigny, SSP; Jerbourg, SM. Hard 'flat' surface.

Platriaux, les — field: la Grande Mare, C. several 'flat areas'. Planitrel q.v.

Plats Pieds, les —

1. house: les Plats Pieds, (le Clos au Comte), C. Origin unknown. House and name date from 1820s.
2. field: les Plats Pieds (les Pièces), F. Origin unknown, recorded **Plas Pies**, 1671.

Platte Hache, **Douit** de la — stream: les Bassièrres, SPP.

Pleinheaume, de — district: Pleinheaume, VdelE. 'flat isle' in a low lying area.

Pleinmont, de — district: Pleinmont, T. 'flat mount, promontory', 'en plein', in open country, ie. in open fields. Recorded de **Plano Monte**, 1289.

Pleinmont, Fief de — feudal holding of land, T. Also known as Fief des Quarante Quartiers.

Pleinmonts, les — field: les Niaux, SA. 'flat mounts'. Recorded, les **Plaidmonts**, 1610.

Plichon, la **Croix** — site: W of les Capelles, SSP. Former wayside cross. Surname extinct 17th c. **Pelichon**. croix q.v.

Plichonnières, les — fields: les Picques, SSV. Recorded les **Pellichonnières**, 1586. Derivation as above.

Plichons, les — fields: W of les Landes du Marché, S of la Grande Maison, VdelE. Derivation as above.

Pluque, de	field: la Palloterie, SPB. Origin unknown.
Pochet, de	field: les Clospains, CduV. Surname extinct 15th c. **Poschet**.
Poidevin(**s**), les	houses, fields, street: le Mont Hermon, Pedvin street, SPP; les Marchez, SPB; la Carrière, F; la Villiaze, les Poidevins, SA. Extant surname **Le Poidevin**. Spelt **Le Pettevin**, **Le Poittevin** until late 18th c. Misspelt as **Pedvin**, as street name in SPP.
Poidevin, le **Clos**	field: la Carrière, F. Derivation as above. clos q.v.
Pointe(**s**), les	houses, fields, all parishes. 'point' of land.
Pointes, **Camp** ès	land: les Grons, ruette Rabey, les Camps, SM.
Pointes, rue des	road: S of Bailiff's Cross Road, SA.
Pointu, le **Courtil**	field: rectory, SM. 'pointed' field.
Pointues Rocques, les	see **Rocques**.
Pois, le **Courtil** à	see **Paix**.
Poisson(**s**), les	see **Paysans**.
Poisson, le **Clos**	land: near les Vallettes, SSV.
Pollet, le	street: N of High street, (la Grande Rue), SPP. Origin unknown.
Pomare, la	houses, fields: la Pomare, SPB. Surname extinct 13th c. **Appemare**. Also feudal holding of land, Fief **Pomare** (**d'Appemare**), SPB.
Pommier, le **Clos** au	house, field: les Rocques, T. Surname extinct 17th c. **Le Pommier**. clos, q.v.
Pommier, la **Hougue** du	house, fields: W of les Forfaitures, C. hougue q.v.
Pommier(**s**), les	field: Woodlands, C.
Pompe, la	house, land: les Mouilpieds. SM. 'pomp', (fine looking).

Ponchez, le — house, land: les Bourgs, C. Origin unknown.

Pont, le — houses, land, fields: la Ramée, SPP; les Closios, CduV; le Pont Renier, le Pont, SPB; Torteval church, T; les Bailleuls, SA. 'bridge' for crossing a stream (douit). planque q.v.

Pont, le **Grand** — harbour: SSP. 'the great bridge'. St Sampson's bridge, a low water crossing now under the seaward pavement alongside the harbour wall, and about the width of that pavement. A hump-back section (submerged at high water) took the roadway over the stream flowing into the harbour at low water.

Pont, rue du — road: unlocated, SSP; le Pont, SPB.

Pont St Michel, le — Clos du Valle, see **St Michel**, le **Pont**.

Pont(s) with attached proper names. See also individual entries.

Pont Allaire, le — see **Allaire**.

Pont du Château des Marais, le — see **Marais**.

Pont ès Clercqs, le — see **Clercqs**.

Pont Hané, le — see **Hané**.

Pont au Meltier, le — see **Meltier**.

Pont Mourinel, le — see **Mourinel**.

Pont Perrin, le — see **Perrin**.

Pont du Petit Marais, le — see **Marais**.

Pont Renier, le — see **Renier**.

Pont Saint Michel, le — see **Saint Michel**.

Pont Vaillant, le — see **Vaillant**.

Pontac, le — former house: le Carrefour, C. Origin unknown. Earliest record Fief de Saumarez, c. 1840.

Pontier, le — field: la Quevillette, SM. Origin unknown.

Ponton, le	field: Airport, F. Origin unknown.
Pontorson, le	bridge: large stream crossing off rue de la Fontaine, SPP, near the Town church. Origin unknown, possibly after similar construction at Pontorson, Normandy.
Porc(s), **Pors**, à	fields: les Genâts, la Porte, Seaview farm, Les Fauquets, C; les Ménages, la Vieille rue, SSV; le Haut Chemin, SPB; le Chêne, F; la Villette, SM; les Blicqs, SA. For 'the pigs'.
Porc, **Camp** au	land: near les Blicqs, SA.
Port au **Vrac**	land: le Pezeril, CduV. 'door', natural entrance for vraic. Recorded **Porte Vracq**, 1591.
Port, de **Bon**	see **Bon Port**.
Port de Fermains, le	see **Fermains**.
Port de Glategny, le	see **Glategny**, **Salerie**.
Port de Nermont, le	see **Noirmont**.
Port de Saint, le	see **Saints**.
Port, du	fields: la Couture, SPP; les Duvaux, SSP. Extant surname **Du Port**. Also feudal holding of land, Bouvée **Catherine Du Port**, C.
Port Grat, le	see **Vaugrat**.
Port, le **Long**	mooring anchorage: N of Côbo. C.
Port, le **Petit**	small inlet: W of Jerbourg, adjacent to le Mur q.v. Originally la Petite Porte, a side entrance to Jerbourg through le Mur, without reference to the inlet now known as le Petit Port.
Port, rue du **Bon**	road: S of la Fosse, SM.
Port Soif	bay, dunes: Les Mielles, VdelE. 'dry, thirsty area'. Recorded **Pour Soif**, contract, 1846.
Port Vase	houses: le Platon, SPP. 'large door', **Porte Vase** set in wall at top of 'way' from les

	Maisons Brulées (Burnt lane) to le **Platon**. Name misspelt as **Port Vase** since 19th c.
Porte(s), la, les	houses, fields: la Grande Rue, SPP; les Grandmaisons, la Hougue à la Perre, les Portes, les Martins, SSP; la Ville Baudu, CduV; les Videclins, la Haye du Puits, la Porte, les Roussiaux, C; la Porte, la Neuve Maison,SSV; les Adams, SPB; l'Epinel, la Corbière, la Villiaze, F; les Mouilpieds, Bon Port, le Mont Durand, SM; les Rohais, SA. 'door' to house or property.
Portelet, **Camp** du	land: la Rocque, T.
Porte, **Camps** au **Sommet** de la **Petite**	land: near le Mur Jerbourg. SM.
Porte, **Clos** de	land: la Neuve Maison, SSV.
Porte du **College**, la	former 'door' to Elizabeth College: at lower part of rue des Frères, SPP.
Porte, le **Douit** de la	stream: les Traudes, SM. Giving 'entrance' through a wide valley to la Bellieuse, q.v.
Portelet, le	house, land, fields: Pleinmont, T; W of Petit Bot, F. Diminutive of port, a small haven. rocque q.v.
Portelettes, les	houses, land, fields: Pleinmont, T. Narrow entrance to summit from E via le Chemin le Roi (present main road).
Portes, les **Grandes**	area of land: les Pezerils, Pleinmont, T. large 'entrance', via le Chemin le Roi to summit of Pleinmont.
Portes, les **Petites**	smaller area of land: S of les Grandes Portes, Pleinmont, T. small 'entrance' through a gap in the rock mass to summit of Pleinmont.
Portier, au	former field: les Rohais de haut, SA. Surname extinct 15th c. **Le Portier**.
Portière, la	houses, land, all parishes. 'garden by the door'.

Portinfer, de	bay: Portinfer, VdelE. 'port of hell', i.e. turbulent seas at all times. Recorded **Portu de Enfer**, 1309. See Enfer.
Potage, à	fields: la Claire Mare, SPB; la Villiaze, F; les Pages, SM; la Villiaze, les Archiers, SA. 'vegetable' patch.
Potel, le **Vau**	fields: le Vaupotel (la Fosse), SPB. Derivation as above. val q.v.
Potellerie, la	fields: le Lorier, SPB; N of Oberlands, SM. Courtil de la Potellerie shortened to Courtil de Paul since 19th c. Surname extinct 15th c. **Postel** (**Pautel**).
Potère, la	land: le Beauvalet, SSV.
Potère, la ruette de la	lane: le Beauvalet to Sous l'Eglise, SSV. Origin unknown, earliest record **Pausterre**, 1580.
Poudreuse, la rue	road, district: les Hamelins, SM. 'dusty, powdery' surface.
Poullain, au	field: les Fortaitures, C. Surname extinct 15th c. **Le Poullain**.
Pouquelaie, la	land, fields: la Couture, les Croûtes, le Foulon, SPP; la Pouquelah, les Houmets (E of Rosemount), C; Plaisance, les Adams, les Reveaux, les Sages, SPB. A late medieval name for 'chamber tombs' based on 'fairies'. None of the above tombs now remain. Also spelt Pouquelah, Pouquelaye, Pouquelée. Déhus, Déhuzel, Creux, Trépied, q.v.
Pouquelaie des **Houmets**, la	field: N of Saumarez park, C.
Pouquelaie des **Marchez**, la	field: les Marchez, SPB.
Pouquelaie des **Rohais**, la	field: le Douit des Rohais, SPP.
Pouquelaie, la rue de la	road: N of le Foulon, SPP.

Pouris, les	field: la Claire Mare, SPB. the 'rotten, putrified'.
Poussin, de	field: les Beaucettes, CduV. Surname extinct 15th c. **Pouachin**. Recorded 1654.
Pré, le	1. field, all parishes, 'meadow'. 2. fields: les Trachieries, SSP; la Croûte, le Gron,SSV; le Pré, les Prés, SPB; Airport, SSV/F/SA; les Prés, SA. 'meadow' as a definite place name.
Pré d'Aval, le	see **Aval**.
Pré Bourdon, le	see **Bourdon**.
Pré de **Murie**, le	see **Murie**.
Pré, le **Grand**	houses, land: la Rochelle, la Hure Mare, CduV; Albecq, C; les Camps. SM. 'big' meadow.
Pré, le **Petit**	house, land: Albecq, C. 'small' meadow.
Pré, le **Rouge**	field: S of les Mielles, CduV. 'reddish' meadow.
Pré Mourin, le	see **Mourin**.
Pré, rue du	road: W of le Bordage, SPP.
Pré Rond, le	land: les Beaucamps, les Genats, C. 'rounded' meadow.
Préel, le	houses, fields: le Préel, les Houmets, C; l'Erée, SPB; les Vaubelets, SA. Surname extinct 14th c. **Du Prael**.
Préel, **Clos** au	land: le Préel, C.
Prés Bourgeois, les	see **Bourgeois**.
Prés des **Bergers**, les	see **Bergers**.
Prestre, le **Croix** au	site: rue de la Cache, C. Former wayside cross. Surname extinct 15th c. **Le Prestre**. croix q.v.
Prestres, Fief des	feudal holding of land. Dependency of Fief de Carteret, C.

Prestres, la rue des	road: Hospital lane, SPP. Former name of lane.
Prêtre, au	land, fields: rue a l'Or, SPP; rue de la Cache, le Marais, C. Surname extinct 15th c. **Le Prestre**. Also feudal holding of land, Fief ès **Prestres**, C.
Prêtre, le **Camp** au	field: le Catillon, SPB. Derivation as above. camp q.v.
Prêtre, la **Chaise** au	site: St Clair, SSP. Former standing stone. Derivation as above. megalith q.v.
Prêtre, la **Hougue** au	field: les Marais, CduV. Derivation as above. hougue q.v.
Prêtre, la rue au	road: E of rue du Presbytère, C.
Prevosts, les	district: les Prevosts, SSV.
Prevost, la **Trappe**	field: les Rocquettes, SSV. Extant surname **Le Prevost**. trappe q.v
Prévôté, la	fields: la Prévôté, SPB; Pleinmont, T. From 'prévôt'. A perquisite enjoyed by the officer (prévôt) of Fief ès Corbinez appointed annually to collect feudal dues. Similarly a perquisite at Pleinmont enjoyed by the prévôt of Fief de Pleinmont.
Prévôté, la **Maison** de **Garde**	watch house: la Prévôté, SPB. Built 1804. garde q.v.
Prévôtiaux, les	land: les Grantez, C. Pieces of land formerly enjoyed by the prévôt of Fief des Grantez.
Prey, de **Jean**	fields: le Grand Hommel (Victoria Avenue), SSP. Surname extinct 18th c. (**Du**) **Prey**. Jempreys q.v.
Priaulx, de	fields: les Arguilliers, le Hamel, VdelE; le Ponchez, C; la Carriere,F. Extant surname **Priaulx**. Misspelt **Prios**, 18th c.
Priaulx, la rue des	road: le Hamel, VdelE. Also called rue des Chapelles. Derivation as above.

Prier, de	land: la Hougue du Pommier, le Préel, C; N of les Bordages, SSV; N of le Chemin le Roi, F. land 'to desire'.
Prier, rue de	road: unlocated, SSV.
Prins(es), les	houses: E of Port Soif, VdelE. After 'Prinse des Mielles' (26 vergées), a feudal holding and dependency of Fief le Comte.
Profonde, la	cliff: Pleinmont, T. the 'depth', i.e. steep.
Profond Chemin, le	see **chemin**.
Profond Courtil, le	see **courtil**.
Profonde Rue, la	see **rue**.
Profonde Rue, la	see **rue**.
Profond Val, le	see **val**.
Profonds Camps, les	see **camps**.
Profonds Camps, les	see **camps**.
Promenades Publiques, les	public walks: l'Hyvreuse, SPP. Now Cambridge park.
Prospect place,	villa: les Buttes, SPB. Built for George Dobrée, 1820s. Now renamed St Bride's.
Provendel, le	field: la Hougue Falle, SPB. 'little provender'.
Provendel, la rue du	land: la Hougue Falle, SPB. Adjacent to above.
Puchée, la	field: les Duvaux, SSP. Origin unknown.
Puits, le(les)	fields, land, all parishes. 'well'.
Puits Binet, le	see **Binel**.
Puits, **Clos** du	land: near la Claire Mare, SPB. Also spelt Puis.
Puits, **Croûte** du	land: le Rond Marais near North side, CduV.
Puits de **Caubo**, **Fontaine**	spring: ruette de la Tour, C.

Puits de Côbo, le	see **Côbo**.
Puits, la **Haye** du	see **Haye de Puits**, la.
Puits, la rue de	road: Well road, Glategny, SPP.
Pulias, de	district: N-W of les Vardes, SSP. Origin unknown, but 'pulies' might mean 'stinking' pond.
Pulias, **Hougue** de	land: near les Vardes, SSP.
Pulias, la **chapelle** de	site: (unlocated) N-W of les Vardes, SSP. Medieval chapel dedicated to Notre Dame. Chaplain owned land, rue des Marais, VdelE, 1470. Possible derivation as above. chapelle q.v.
Putron, De	fields, land, district: les Terres, Fort George, SPP; les Mouilpieds, la Marette, la Ruette, SM. Extant surname, **De Putron** (**De Puteron**).
Putron, **Camp** de	land: la Ruette, W of la Marette, SM.
Putron, **Douit** de	stream: near Fort Geroge, SPP.
Putron, la **Croûte** de	field: Fermains, SPP. Derivation as above. croûte q.v.
Putron, la **Mare** de	field: Fort road, SPP. Derivation as above. mare q.v.
Putronne, la	field: la Marette, SM. Derivation as above.
Putronnerie, la	see **Pitronnerie**.

Q

Quai, le	field: W of la Ville Baudu and les Mares Pellées, CduV. Former quay, le Cay, 1654. The remains of the medieval quay run the entire length of the land bordering the Braye du Valle between la Ville Baudu and les Mares Pellées.

Quaingo(t), le — field: les Paysans, SPB. Origin unknown. le **Quaignot**, 1671.

Quanteraine, de —
1. houses, fields: rue des Marottes, C. Recorded **Cantereine**, **Canterayne**, 1470. **Quanteraine**, 1634.
2. house, fields, water mill: N of les Sages,SPB. moulin à eau, Cantereine, q.v. Also feudal holding of land, Fief de **Quanteraine**, (Cantereine) SPB/SSV. Mid-late 12th c. grant by seigneur of Fief au Canely q.v. to Abbaye de Notre Dame du Voeu, Cherbourg. Known as Chantereyne or Cantereine because of the croaking frogs in the nearby marsh.

Quanteraine, rue de — road: W of la Hougue Falle, SPB.

Quarantaine, la — land: la Croix Bertrand, SM. Refers to a rente of 40 eggs. see Carantaine.

Quarante Quartiers, Fief des — see Fief de **Pleinmont**, T.

Quart d'Ecu, de — field: le Foulon, SPP. A 'quarter crown', i.e. value of a rente due on that field.

Quartier(s), les — houses, fields: les Quartiers, SSP; le Carrefour, C. Medieval surname **Cartier** (**Caretier**), extant and spelt since 16th c. **Quertier**.

Quartier du Camp — feudal holding of land. Dependency of Fief Rozel, VdelE.

Quartier, le **Clos** — land: le Carrefour, C. Derivation as above. clos q.v.

Quartiers, **Hougue** des — land: E of les Osmonds, SPP,

Quartiers, route des — road: W of Route de Coutanchez, SSP.

Quathoule, la — land: les Pointues Rocques, SSP. 'cavity'.

Quatière, la — house, land: les Genâts, C; Pleinmont, T. Origin unknown. Spelt Caquinère, 1873.

Quatre Chemins, les — see **Chemins**.

Quatre Vents, les — see **Vents**.

Quauquaire, la	land: la Villiaze, SA. Recorded 1745. Origin unknown.
Quemin, le **Bas**	see **Chemin**, le **Bas**.
Quence, de	field: Roulias, F. Origin unknown.
Quenot, le **Hommet**	see **Hommet**. Surname extinct 16th c. Quenot.
Quenots, les	field: les Sauvagées, SSP. Derivation as above.
Quensue, de	field: Baugy, CduV. 'no sweat'. Recorded: 'le courtil **que ne sue'**, 1591.
Querette, le **Clos**	field: la Ville Amphrey, SM. Surname extinct 14th c. **La Carette**. Recorded **Queruette**, 1721, **Reymundus La Carette**, 1309. clos q.v.
Queripel, de	land: les Monts, SSP; la Vallée, SPB; Airport, F; la rue au Cammus, SM; les Eperons, la Villiaze, le Grand Belle, SA. Extant surname **Queripel**, until 17th c. spelt **Carupel**.
Queripel, de **Thomas**	land: le Grand Belle, SA. Derivation as above.
Queritez, les	houses, fields: le Camp, les Delisles, C. After **Thomasse Caritey**, 1470. Surname extinct 16th c. **Caritey**.
Querpentier, au	land: les Rohais, SPP. Surname **Le Carpentier** (**Le Querpentier**).
Querpenterie, la	fields: les Monmains (Monts Mains), la Hougue Patris, CduV. Relates to the above surname **Le Carpentier** and also to land not owing chefrente.
Querpentiers, la **Mare** ès	pond: S of Hacsé, CduV. Derivation as above. mare q.v.
Querqueux, le	cliff: S of le Crolier, T. Origin unknown. Recorded 1787.
Querrière, la	fields: les Houmets, le Marais, C; rue de l'Arquet, SSV; Pleinmont, T; la Fallaize, SM.

	Permanent 'stony' cart track, usually a legal servitude as under **Charrière**. cache, charrière, traversain q.v.
Querrière, la **Neuve**	area of land: N of le Catioroc, SSV. 'new'. Recorded 1536. See above.
Querrier, la rue au	main road: W of Castel church, C. 'stony'. Recorded 17th c.
Querruée, la	see **Charruée**, la.
Querterie, la	see **Charterie**, la.
Quertier, de	house, field: les Pelleys, rue Cohu, C; les Jaonnets, SSV; les Fauconnaires, SA. Extant surname **Quertier**, medieval spelling **Cartier**, **Quartier**(**s**) q.v.
Quertière, la **Brèche**	land: rue de l'Arquet, SBP; les Portelettes, T. Derived from above surname **Quertier**. brèche q.v.
Quesnel, la **Croûte**	field: la Hougue Juas, CduV. Extant surname **Quesnel**. croûte q.v.
Quetteville, **De**	house, field: St Jacques, la Vrangue, SPP. Surname extinct 18th c. **De Quetteville**.
Queue de **Rat**, le	fields: N of les Jetteries, SSV. 'rat's tail', i.e. at far end. Recorded 'de **queuée de ratt'** 1695.
Queux, les	house, fields: W of les Effards, C. After **Perrotyne Le Queu**, 1548. Surname extinct 17th c. **Le Queu**. Also feudal holding of land, Fief des **Queux**, C.
Queux, la **rue** des	road: les Queux, C. Derivation as above.
Queux, le **Val** au	land: les Terres, SPP. Derivation as above.
Quevillette, la	fields, road: W of les Vaurioufs, SM. Origin unknown.
Quevillons, les	fields: les Etibots, Seaview farm, C. Surname extinct 15th c. **Quevillon**.
Qui tempête, la rue	see **Tempête**, la rue qui.

Quite, la	field: les Arguilliers, VdelE. Origin unknown. Earliest ref. 1838.
Quittempois, de	field: les Fontaines, C. 'depart in peace' (quitte en paix) i.e. derived from a rente for a requiem.

R

Rabé, de	field: la Pouquelah, C. After **Daniel Rabé**, 1734.
Rabey, de	fields: la Fosse, le Rocher, SM. Extant surname **Rabey**.
Rabey, **Ruette**	lane: W of la Ville Amphrey, SM.
Rachel, de	field: Fort George, SPP. Forename (f) **Rachel**.
Raie, la **Long**(**ue**)	houses, fields: la Couture, SPP; E of l'Ancresse, la Hure Mare, CduV; le Tertre, C; S of les Hougues mill, SSV; Airport, F. 'long baulk, furrow' of ground separating strips of land (camps de terre) within a 'shot' (riage). Recorded **Longue Raye**, 1591.
Rais, les **Rouges**	field: les Baissieres, SA. 'reddish baulk' of ground. Also spelt **Rais**, **Raye**, **Rée**. camp, riage q.v.
Raies, les	houses, land: les Bordages, SSV; les Buttes, SPB. Extant surname **Le Ray**.
Raies, rue des	road: E of les Bordages, SSV.
Raineterie, la	see **Reineterie**, la.
Rainsions, les	fields: N of le Courtil Ronchin, SSV. Possibly 'grooved furrow' on a slope.
Ralles, rue ès	land: N of la Moye, CduV. Origin unknown, but may be linked to **Raulle**, an animal 'pound', see below.

Ramée, la	district, road: la Ramée, SPP. Surname extinct 15th c. **De La Ramée**. croix q.v.
Rameurs, les	house, land: now called Home Farm, les Houmets, C. Surname extinct 16th c. **Rameur**.
Rameurs, rue des	road: unlocated, C.
Rauf, **Fontaine**	spring: les Marettes, SM.
Rauf Salemon, **Croûte**	land: unlocated, CduV.
Raulin, de	garden: la Couture, SPP. Forename (m) **Raulin**.
Raulle, la	land: S of le Variouf overlooking valley, F. Probably a 'pound', being walled to a significant height with direct access from tracks N and S.
Ravenios, les	land: N of Saumarez park, C. Surname extinct 15th c. **Ravenel**. croûte q.v.
Ravenios, **Croûte** des	land: N of Saumarez Park, C.
Ravine, la	field: le Douit, SSV. 'torrent'.
Ravines, les **Grandes**	land: le Douit, SSV.
Ray, au	house, field: le Vaugrat, SPP; Pleinheaume, VdelE. Extant surname **Le Ray**.
Ray, le **Ménage Josué Le**	field: new cemetery, le Long Frie, SPB. Formerly house of **Josué Le Ray**, 18th c. Derivation as above. ménage q.v.
Ray, de **Thomas Le**	field: la Croix Brehaut, SPB. Derivation as above.
Rébouquets, les	house, fields: S of Forest road, F, SM. 'refilled'.
Rebours, le **Bordage**	feudal holding of land, unloc. CduV. Bordage q.v.
Reculet, le	house, fields: la Fosse, SM. 'remote'.
Rée, la **Long**(ue)	see **Raie**, la **Longue**.

Rée, de le	see **Erée**, l'. 19th / 20th c. misspelling of surname **De Lerée**.
Reine(**s**), la(les)	houses, fields, district: la Bernauderie, les Videclins, C; les Reines, on SPB/F parish boundary; les Hubits, SM. 'green frog', i.e. a marshy area.
Reines, rue des	road: W of Rue des Farras, F.
Renault, de	house, fields: les Rouvets, SSV; Bon Port, SM. Extant surname **Renault**, also spelt **Renaud**. croûte q.v.
Renault, **Croûte**	land: Bon Port, SM.
Renault, la **Vallée**	land: les Cambrées, SPB.
Renier, le **Pont**	house: les Ruettes Brayes, SPP; les Marchez, SPB. 19th c name for house, originally called le Pont.
Renonciation, la	fields: Haut Sejour, C. 'renunciation of title', by John Cohu to H.M. Receiver, 1470.
Renost, le **Bordage**	feudal holding of land, unloc. CduV. Bordage q.v.
Renouards, les	former fields: le Feugré, C. Surname extinct 15th c. **Renouard** (extant Jersey).
Renouf, de **Jean**	field: la Palloterie, SPB. Extant surname **Renouf**.
Renouf, la **Hougue**	hillock, fields: St Germain, C, Derivation as above.
Renouvet, de	land: les Terres, SPP. Forename (m) **Renouvet**, meaning 'renewed', i.e. born again.
Renouvet Hamelin, **Croûte**	land: W of the chapel of St Magloire, CduV.
Répos, Bon	see **Bon Répos**.
Reserson, de	garden, field: Cornet street, SPP; le Pont Perrin, VdelE. Surname extinct 19th c. **Reserson**.

Rétot, de — house, fields: Sous les Courtils, C. Surname extinct 16th c. **Restault**. Also mis-spelt **Rétos**.

Reveaux, **Franc Fief** des — feudal holding of land, SPB / SSV. Revel q.v.

Revel, de — fields: les Sages, la Hougue Falle, rue St Pierre, SPB. Surname extinct 16th c. **Revel**. Also feudal holding of land, Franc Fief des **Reveaux**, SSV/SPB.

Revel, de **Collas** — field: le Marais, C. After **Collas Revel**, 1536.

Revel, de **Dan Nicolas** — see **Croix** de **Dan Nicolas Revel**.

Revel, de **Simon** — fields: Pleinmont, T. After **Simon Revel**, 1549.

Revers, de — fields: le Grande Lande, SSV. 'back' field.

Riage, le — land: St George, la Masse, la Hougue au Comte, C; le Felconte, la Hougue Anthan, SPB; les Camps, SM. 'shot' or run of strips of land (camps de terre) in one entity or field, in slightly curved form to aid use of a narrow headland (buttière de terre) for turning to enter the next furrow (sillon) at right angles. buttière, camp, raie, sillon q.v.

Ribet, de — fields: les Etibots, VdelE; Reservoir, les Choffins, SSV; les Caches, SM; les Mourants, SA. Surname extinct 17th c. **Ribet**. Also misspelt **Ribault**, **Ribées**, **Ribert**, **Ribot**.

Ribonne, la — field: rue du Clos, SSP. Origin unknown.

Ricart, de — fields: la Folie, les Mielles, CduV. Surname extinct 15th c. **Ricart**.

Ricart, **Cour** — fields: les Clercs, SPB. Derivation as above.

Ricart, **Fontaine** — spring: Jerbourg, SM.

Ricart, la **Hougue** — houses: les Loriers, CduV. Derivation as above.

Ricart, les **Messières** — land: Jerbourg, SM. Derivation as above. hougue, messières q.v.

Richard, **Clos**	land: near le Bourg, F.
Richard, de	field: les Gigands, SSP.
Richard, de **Jean**	field: E of la Lande, C. After John Richard, 1470. Surname extinct 16th c. **Richard**.
Richard de la Folie, Fief de	feudal holding of land. Dependency of Fief de Rozel, SSP.
Richard de Nermont, Fief de	feudal holding of land. Dependency of Fief de Rozel, VdelE.
Richard des Camps, Fief de	feudal holding of land. Dependency of Fief de Blanchelande, SM.
Richarderie, la	field: les Fontaines, T. After **Richard Galllienne**, 1775.
Richez, la **Ville** ès	see **Ville** ès **Riches**.
Richmond	district: les Jenemies, SSV. Named after Fort Richmond. Temperance hotel, 19th c., now commercial premises, also took the same name.
Richmond Corner	road junction: les Bas Courtils, Bulwer avenue, SSP. Called after 19th c. house on corner.
Richmond, **Fort**	former barracks, battery: le Mont au Nouvel (q.v.), SSV. Built mid-19th c.
Ride, de	field: les Massies, SSV. After **James Ride**, 17th c.
Rideaux, à	field: la Fallaize, SM. Possibly from extinct surname **Ride**.
Riebour, de	field: N of Bordeaux, CduV. Surname extinct 15th c. **Rebours**. Also spelt **Riquebourcy**. Also feudal holding of land, Bordage **Rebours**, CduV.
Rigeaux, les	field: les Lohiers, SSV. Origin unknown.
Rignet, le	fields: la Maladerie, SSV. Surname extinct 16th c. **Riguenet**, spelling contracted in the 20th c.

Rihouel, de — field: les Landes, CduV. Extant surname **Rihoy**, 19th & 20th century spelling.

Rimonel, le — field: la Lague, SPB. Surname extinct 15th c. **Rimonel**. Also spelt **Rimonet**.

Riolé, de — field: les Mares Pellées, CduV. Surname extinct 15th c. **Le Riolley**.

Riolet, de — field: le Mont Saint, C. After **Collas Le Riolley**, 1470. Derivation as above.

Riollées, de — field: les Rohais, SA. Derivation as above. Also feudal holding of land, Fief des **Riollais**, C.

Rive, la — coast above high water spring tides, all parishes. Also called le **Rivage**.

Rivel, de — strips of land (camps de terre): opposite la Mare, Pleinmont, T. Origin unknown, possibly extinct surname. camp q.v.

Rivière(**s**), le(les) — fields: N of la Vrangue, SPP; Baubigny, SSP; S of les Mielles, CduV; le Marais, C; la Claire Mare, la Rousse Mare, SPB. a 'wide stream'. Also feudal holding of land, Fief de la **Rivière**, C. Avisse, Fief Dame q.v.

Road — see **chemin**, **route**, **rue**, **ruette**.

Road, **New** — road: S of St Sampson's harbour, SSP. Enlarged and widened 19th c.

Road, **New** — road: N of Castel Hospital, C. Built 20th c.

Road, **New** — road: les Bruliaux, SPB, to la Villiaze, F. Rebuilt early 19th c.

Roads, — English usage, roads with proper names, omitting 20th c. road signing.

SPP — **Amherst** road, **Battle** lane, **Belmont** road, **Brock** road, **Candie** road, **Colborne** road, **Collings** road, **Dalgairns** road, **Doyle** road, **Forest** lane, **George** road, **King's** road, **Paris** road, **Prince Albert** road, **Queen's** road, **St Julian's** avenue, **Victoria** road.

SSP — **Brock** road, **Bulwer** avenue, **Commercial**

	road, **Victoria** avenue. street, chemin, rue, ruette q.v.
Roberge, de	fields: Belval, Bordeaux, Sohier, CduV; E of le Bigard, F. Surname extinct late 20th c. **Roberge**.
Roberge, **Ruette** de la	land: le Bigard, F.
Robergerie, la	district: N of St Clair, SSP. Derivation as above.
Robert(**s**), les	fields: les Domaines, SSV; l'Alleur, T; Forval, le Bourg, F. Extant surname **Robert**.
Robert Allez, la **Fosse**	land: les Fosses, F.
Robert de Vic, Fief de	feudal holding of land, SPB / T. Also known as Robert de Verre.
Robert, **d'Elie**	field: les Niaux, C. Extant surname **Robert**.
Robert, de **Jean**	fields: les Jaonnets, C; le Grais, les Buttes, SPB. Derivation as above.
Robert, la **Croûte**	field: Pleinmont, T. Derivation as above.
Robert, le **Val**	field: le Mont Hérault, SPB. Derivation as above.
Robilliard, de	fields: la Pointe, les Brehauts, SPB; la Forge, T; les Nicolles, F. Extant surname, **Robilliard**. Spelt **Robillard**, **Robilliart**, before 1750.
Robin, de	fields: les Longs Camps, SSP; les Landes, Saumarez park, C; S of les Sablons, SPB; la Colombelle, Bon Port, SM; les Eperons, Ste. Hèlène, SA. Extant surname and forename (m) **Robin**.
Robin, de **Giret**	field: les Vardes, SSP. Derivation as above.
Robin, de **Jean**	field: la Pomare, SPB. Derivation as above.
Robin, la **Croix**	large site at road junction: le Coudré, SPB. wayside cross. Latest date mid 13th c. from feudal evidence, re Fief au Canely. Derivation as above.

Robin, la **Fontaine**	spring: les Maindonnaux, SM.
Robin, la **Hougue**	field: les Français, CduV. Recorded since 1879. Derivation as above.
Robin, rue	fields: Bordeaux, CduV. Derivation as above.
Robine, la	house, land: Cobo, C. Feminine version of **Robin**.
Robinerie, la	house: la Couperderie, SPP. See above surname **Robin**.
Robinet, de	land: les Camps. SM. 'pet' form of **Robin**, forename or surname. verbeuse q.v.
Robinette, la	house, land: Cobo, C. 'pet form' of **Robine**.
Robin Sohier, de	see **Sohier**.
Roches, les	field: les Rébouquets, SM. 'rocks'. rocque q.v.
Rochelle, la	former field, district: Vale road, SSP; E of la Moye, CduV. A 'small rock mass', first recorded 1654.
Rocher, le	house, land: les Annevilles, SSP; S of le Gelé, C; les Fontaines, T; S of la Fosse, SM. 'the rock'. rocque, roquer q.v.
Rocher (**Roquer**), rue du	road (lane): S of le Gelé, C.
Rocheuse, la rue	see **rue Rocheuse**.
Rocquage, **Fontaine**	spring: Jerbourg, SM.
Rocque à Beuf, la	St George, C. Beuf q.v.
Rocque à Gelline, la	les Pages, SM. 'chicken rock'.
Rocque à l'**Ane**, la	fields: les Terres, SPP; S of le Marais, SSP. 'rough rock outcrop', 'donkey rock'.
Rocque à l'Eau, la	les Choffins, C. eau q.v.
Rocque à l'Or, la	N of rue à l'Or, SPP; Pleinmont, T. 'golden rock'.
Rocque au Cheval, la	Fort George, SPP. Cheval q.v.

Rocque au Follet, la	la Rocque, la Hougue Anthan, SPB. Follet q.v.
Rocque au Mer, la	N of Côbo (rue des Carterets) C. 'sea rock'.
Rocque au Sermonier	l'Erée, SPB. Sermonier q.v.
Rocque au Tonnerre la	le Catioroc, SSV. 'thunderbolt rock'.
Rocque au Vère, la	Bon Port, SM. Vère q.v.
Rocque aux Fées, la	le Bourg, F. Fées q.v.
Rocque aux Français la	la Hougue du Pommier, C. Français q.v.
Rocque aux Paints, la	la Soucique, F. Paint q.v.
Rocque Balan, la	S of l'Ancresse, CduV. Balan q.v.
Rocque Beaucamp, la	les Maingys, VdelE. Beaucamp q.v.
Rocque Chardo, la	W of le Tertre, CduV. Chardo q.v.
Rocque Colin, la	le Gains, SPB. Colin q.v.
Rocque Crespel, la	le Coudré, SPB. Crespel q.v.
Rocque Damalis, la	Icart, SM. Damalis q.v.
Rocque de la Hougue Anthan, la	la Hougue Anthan, SPB. Anthan q.v.
Rocque des Chèvres la	Fort George, SPP. Chèvres q.v.
Rocque du Portelet, la	Pleinmont, T. Portelet q.v.
Rocque du Val, la	S of Rocquaine, SPB. val q.v.
Rocque ès Cas, la	N of la Moye, CduV. Cas q.v.
Rocque Fendue, la	le Catioroc, SSV. 'cleft rock'.
Rocque Galliot, la	l'Ancresse, CduV. Galliot q.v.
Rocque Gohier, la	le Marais, C. Gohier q.v.
Rocque Hamelin, la	rue Poudreuse, SM. Hamelin q.v.
Rocque où le Coq chante, la	le Dos d'Ane, C. Coq, q.v.

Rocque Pendante, la	les Hamelins, SPB. 'steep sloping rock'.
Rocque Poisson (Payson), la	S of rue Rocheuse, SPB. Paysans q.v.
Rocque Poisson, neuf **Clos** de la	land: Rocque Poisson, SPB.
Rocque qui Sonne, la	Vale School, CduV. 'rock which rings'.
Rocque Tourgis, la	la Bouvée, SM. Tourgis q.v.
Rocque, la **Blanche**	rock: No. 6 berth, harbour, SPP. 'white rock'. Probably quartz or mica outcrop. blanche q.v.
Rocque, la **Grosse**	land: W of la Mare de Carteret, C; la Hougue Fouque, SSV; l'Erée, le Felconte, SPB; Pleinmont, T; la Villiaze, SA. A 'fat rock'.
Rocque, la **Longue**	land, fields: la Couture, les Croûtes, les Rohais, S of la Ramée, SPP; les Trachieries, SSP; E of les Landes, CduV; la Marette, SSV; N of le Felconte, les Paysans, SPB. 'standing stone'. Stones at la Marette and les Paysans survive. pierre, longue, q.v.
Rocque, la **Platte**	field: S of le Crolier, T. 'flat rock'.
Rocque, la **Ronde**	land: Icart, SM. 'round rock'.
Rocque(s), les **Blanches**	houses, fields: la Masse, les Houmets, S of le Villocq, C; W of le Catillon SPB; la Soucique, F. blanche q.v.
Rocque, **Rocquelin**	land: les Hougues, C. Origin unknown. First recorded as **Herclin Roche**, 1179. vivier, le clos q.v.
Rocquer, le	see **Rocher**, le.
Rocquer, rue du	see **Rocher**, le.
Rocquerel, le	see **Rocré**, le.
Rocques aux Caux, les	S-E of Rocquaine, SPB. Cauf q.v.
Rocques Barrées, les	S of Bordeaux, CduV. 'barred rocks'.

Rocques de Totenet la	les Rocques, T. Totenet q.v.
Rocques Fouchin, la	le Campère, SPB. Fouchin q.v.
Rocques, le **Clos** des **Sept**	field: les Vardes, SSP. 'seven standing stones'.
Rocques, les **Buis**	fields: Rue Rocheuse, SPB. 'rock shaft'. Recorded, de **Buroque**, 1671.
Rocques, les **Grandes**	land: les Grandes Rocques, C; les Fontenelles, F. 'big rocks, rock mass'.
Rocques, les **Pointues**	site: W of Delancey park, SSP. 'pointed rock'.
Rocquette(**s**), les	houses, fields, all parishes. Diminutive of rocque q.v.
Rocquettes, rue des	les Rocquettes, SPP.
Rocquettes, **Ruette** des	land: E of les Rocquettes, SPP.
Rocqueuse, la rue	see **Rue Rocheuse** (**Rocqueuse**), la.
Rocqueux, **Camp**	land: unlocated, T.
Rocré, le	house, land: Sous les Courtils, C. 19th c. contraction of **rocquerel**. 'small' rock, stone. Diminutive of rocque q.v.
Roque(**s**), le(les)	'standing stone'. chaise, perron, pierre q.v.
Roger Gosselin, Fief de	feudal holding of land. Dependency of Fief de Rozel, VdelE.
Rogier le Villain, Fief de	feudal holding of land. Dependency of Fief de Blanchelande, SM.
Rognons, les	field: les Longs Camps, SSP. Origin unknown.
Rohais, les	houses, fields, land: les Rohais, SPP; les Rohais, SA. Surname extinct 14th c. **De Rohais**. croûte, pouquelaie q.v. Also feudal holding of land, Fief des **Rohais**, SA.
Rohais, **Croûte** des	land: les Rohais, SA.

Rohais, **Douit** des — stream: lower Rohais, SPP.

Roi, le — field: les Blicqs, SA. Sometime in possession of the Crown, probably for non-payment of rentes.

Roi, le **Camp** du — former field: to W of (Camp du Roi) road junction. Centre of former market at les Landes du Marché, q.v. Sold by the Crown in mid 19th c.

Roi, le **Chemin** le — the 'king's way': Traverses the Island.

Start – Vale church, E to le Grand Pont, St Sampson, St Peter Port church, rue Berthelot, la Petite Marche, Jerbourg, le Chemin le Roi, le Bourg, New Road, la Paloterie, la Croisée, la Hougue Anthan, les Portelettes, Pleinmont, Table des Pions, les Sablons, Mont Chinchon, le Mont Saint, les Jenemies, le Dos d'Ane, les Grands Moulins, les Landes du Marché, le Camp du Roi, rue Sauvage, l'Islet, Pont St Michel, Vale church.

Used at regular intervals by the courts of the Crown and Fief St Michel for a joint inspection of the 'highways' around the island, called la Chevauchée St Michel, which functioned between the 12th and 19th centuries. Its origin may have followed the unification of the Bessin and Cotentin halves of the Island under Geoffrey Plantagenet, circa 1140–45.

Roi, Fief le — feudal holding of land by the crown, formerly the major part of the fief of the Vicomte du Cotentin, seized at the time of the Plantagenet conquest of the Island, mid 12th c. Also name of house, les Varendes, SA.

Roi, la **Hougue** du — fields: N of le Monnaie, SA. Probably in possession of the Crown at some time for non-payment of rentes, or from the surname **Le Roy**.

Roi, la **Ville** au	see **Ville au Roi**.
Roi, les **Géans** du	land: Pleinmont, T. strips of gorse alongside 'le Chemin le Roi' . Name recorded in 18th c. les **Jaons** du **Roi**. (**du Roy**)
Roland, de	road, field: rue Roland, SSP; St George, C. Surname extinct 17th c. **Rollant**, **Rolland**.
Roland, de **Thomas**	field: N of rue des Pointes, SA. Derivation as above.
Rolle, le	see **Raulle**, le.
Romains, **Hougue** des	land: les Trachieries, SSP.
Romains, les	house, land: S of les Trachieries, SSP. Surname extinct 15th c. **Romain**.
Romil, le **Clos** au	field: S of la Grand Mare, C. Surname extinct 15th c. **Le Romy**, **Romye**, **Remy**, **Roumy**. clos q.v.
Rompains, les	field: la Hougue Patris, CduV. Origin unknown. Recorded les **Rondspains**, 1654.
Rompeurs, les	former field: N of rue des Carterets, C. Surname extinct 15th c. **Le Rompeur**. Also feudal holding of land, Fief des **Rompeurs**, C.
Roncefer, de	see **Rungefer**.
Ronces, à	fields: la Ronde Cheminée, les Touillets, C. 'brambles,briars'. clos q.v.
Ronces, **Clos** à	land: near les Touillets, C.
Ronces, la rue à	road: E of Galaad chapel, C. Derivation as above.
Ronceval, de	house: rue d'Aval, SSP. 19th c. villa.
Ronchin, **Croûte**	land: unlocated, SM.
Ronchin, de	houses, fields: courtil Ronchin, SSV/SA; les Adams, SPB; les Pièces, F. Surname extinct 16th c. **Ronchin**. croûte q.v.
Rond Camps, les	see **Camps**, les **Ronds**.

Rond Marais	see **Marais**, le **Rond**.
Rond Pays, le	see **Pays**, le **Rond**.
Rond Val, le	see **Val**, le **Rond**.
Ronde Cheminée, la	see **Cheminée**, la **Ronde**.
Ronde Croûte, la	see **Croûte**, la **Ronde**.
Ronde Pièce, la	see **Pièce**, la **Ronde**.
Rondel, de	field: Bordeaux, CduV. 19th c. misreading for extant surname **Blondel**, q.v.
Rondeval, de	field: les Traudes, SM. 'round valley'. 19th c. misreading of le **Rond Val**, q.v.
Rondiaux, les	house, land: S of les Brehauts, SPB. Plural of extant surname **Rondel**.
Rondin, au	fields: les Massies, SSV; les Sages, SPB; l'Epinel, F. Surname extinct, **Le Rondin**. **Estienne Panthon** dit **le Rondin**, 1433.
Rongefer, le	see **Rungefer**.
Roquaze, la	land: W coast of Jerbourg, SM. 'rocky'.
Roque, la	see **Rocque**, la.
Ros, la **Fontaine**	spring: rue des Pages near la Fallaize, SM.
Ros, le **Vert**	fields: N of la Vrangue, SPP; Saumarez park, S of les Français, C. 'green reed'. Also spelt **Rost**.
Rose, de	field: les Prevosts, SSV. From forename (m) of former owner. Recorded **Rosse David**, 1586.
Rose, de	field: l'Echelle, SA. Either forename (m), or surname **Rosse**, **Rose**.
Rosette, le **Ménage**	field: les Villets, F. 'small area of reed'. ménage q.v.
Rosière, **Franc Fief** de la	feudal holding of land, SSP.
Rossiaux, les	land: S of la Maison d'Aval, VdelE. plural of **ros**, 'reeds'.

Rost, **Rot**(**s**), le	see **Ros**, le.
Roteur(**s**), au	land: les Terres, SPP; N of la Mare, le Crocq, SPB; la Bellée, la Ville Tolle, T; S of le Vallon, la Fallaize, SM; les Maumarquis, les Norgiots, les Gouies, SA. 'retting pit' for flax. Also spelt **Routeur(s)**.
Rotière, la	land: les Marais, SPP/SSP; les Nicolles, l'Islet, SSP; la Maraitaine, CduV. 'reed bed'.
Roton, de **Collas**	field: E of route de Sausmarez, SM. Surname extinct 15th c. **Roton**.
Rots, les **Camp**(**s**)	cliffland: S of le Bigard, F. 'strips of land in reed'. Also spelt les **Camprots**. camp, vaux hanniet q.v.
Rots, les **Hougues**	land, hillock: N of les Prins(es), VdelE. Either above derivation or an extinct surname.
Roty, de **Jannin**	field: N of rue Cohu, C. Surname extinct 15th c. **Roty**.
Rouelle, la	field: Airport, F. 'round'.
Rouette, la	field: les Rohais, SA. 'round'.
Rouf, **Camp** du	land: unlocated, VdelE.
Rouge Fallaize, la	see **Fallaize**, la **Rouge**.
Rouge Huis	house: Brock road, SPP. 19th c name.
Rouge, la **Croûte** à la	field: la Maraive, CduV. 'reddish, crusty soil'. croûte q.v.
Rouge Pré, le	see **Pré**, le **Rouge**.
Rougerais, les	see **Rais** (**Rayes**), les **Rouges**.
Rouge Rue, la	see **Rue**, la **Rouge**.
Rouges Petris, les	see **Petris**, les **Rouges**.
Rouget, le **Camp**	field: Chouet, CduV. Derivation as above.
Rougetel, de	field: la Villette, SM. Probably from extinct surname **Le Rougetel**.

Rouget(s), ès	fields: route de Plaisance, les Ménages, SPB; la Ruette SM. Also recorded as Rougier. Surname **Roger**, or **Rouger**, now extant as **Rouget, Rougier**. Also feudal holding of land, Fief au **Rougier** (au **Rouger**) 1723, au **Rogueur**, 1331.
Rouget, le **Camp**	field: Chouet, CduV. Derivation as above.
Rougeval, de	houses, land: Rouge Rue, SPP; W of rue de Laitte, T. 'red valley' from reddish ground, leached on steep slopes.
Rougeval, Fontaine	spring: rue du Rougeval. T.
Rougier, de	fields: les Ménages, SPB; la Ruette, SM. Extant surname **Rouget, Rougier**.
Rougier, les **Vaux**	fields: les Bordages, SSV. Derivation as above.
Rougiers, les **Camps**	field: la Pomare, SPB. Derivation as above.
Roulias, de	fields: N of le Chêne, F/SA. 'reddish, rusty'. Spelt **Rouillais**, 1616, **Rouillayes**, 1671, 1744. Current spelling is 19th c.
Roulleurs, les	land: les Hunguets, SA. Origin unknown.
Roussa(i)llerie, la	house, land: S of la Ramée, SPP. Extant surname **Roussel**.
Rousse, la	land, field, battery: Hacsé, CduV; les Dicqs, VdelE. 'reddish'. Martello tower No.11 q.v.
Rousse Mare, la	see **Mare**, la **Rousse**.
Roussiaux, les	house, land: les Pelleys, C. Extant surname **Roussel**. Owned by family, **John Roussel** 1555, until **Nicolas Roussel**,1853.
Roussillons, les	fields: rue Rocheuse, SPB. 'reddish furrows'.
Route, la	house, land: Rocque au Mer, W of rue des Carterets, C; route de Plaisance, SPB. 'main road'. Both houses have since been renamed.

Route, la — 'main road', all parishes. Listed under a proper name in the main text. It appears to be a haphazard description, as many 'routes' are minor roads.

Rouvets, les — houses, fields: les Rouvets, VdelE; les Rouvets, SSV. Surname extinct 16th c. (**Le**) **Rouf**. Also spelt les **Rouvés**, les **Rouvez**. croûte q.v. Also feudal holding of land, Fief du **Camp Rouf**, VdelE.

Roux, de — fields: la Fosse, SPB; le Manoir, F. Surname extinct 16th c. **Le Roux**. Recorded **Rauf**, 1309. Also feudal holding of land, Fief au **Roux** (Rouf), SSV.

Rozel, de — houses, fields: Fort George, le Mont Durand, rue Rozel, SPP; les Grands Moulins, C; la Quevillette, SM; les Huriaux, SA. Surname extinct 17th c. **De Rozel**. rue q.v. Also two separate feudal holdings of land, Fief de **Rozel**, SPP, and Fief de **Rozel**, VdelE.

Ruaux au **Beir**, les — land: le Dos d'Ane, C. Derivation as above. Beir, le q.v.

Ruaux, les — fields: les Duvaux, SSP; field, N of les Laurens, T. Plural of ruel, 'stream'. Also spelt **Ruos, Rues**.

Ruel, le — land: le Bouet, les Terres, N of Fermain, rue de la Fontaine, Glategny, SPP; la Rousse Mare, SPB; Icart, SM. 'minor stream' or runnel'.

Rue d'**Aval**, la — roads: les Hougues, SSP; N of le Coudré, SPB. 'sloping' road.

Rue de **Derrière**, la — road: la Contrée Mansell, SPP. 'road at back', i.e. Back street.

Rue de **l'Eglise**, la — see **chemin de l'Eglise**.

Rue des **Grandes Rues**, la — road: les Reveaux, SPB; 'road between main roads' i.e. connecting the main road, formerly a narrow lane between les Reveaux and les Islets.

Rue, dessous la — fields: Bordeaux, Beaucette, CduV. Fields 'below' the road.

Rue, dessus la — field: le Russel, F. Field 'above' the road.

Rue Frairie, Fief de la — feudal holding of land. SA.

Rue, la **Droite** — field: la Hougue Fouque, SSV. The 'straight' road.

Rue, la **Grande** — houses, land: la Grande Rue (High Street), SPP; N of Baugy, E of les Grippios, CduV; la Grande Rue, SSV; la Grande Rue, SM. 'main road'.

Rue, la **Longue** — houses, fields: le Foulon, SPP; les Landes du Marché, VdelE; S of le Préel, middle of la Grande Mare, C; S of les Jenemies, SSV; S of les Caches, SM. 'long' straight road.

Rue, la **Neuve** — houses, land: la Ramée, la Couture, Havelet, SPP; N of les Duvaux, SSP; rue de Calais, SM. 'new' road.

Rue, la **Profonde** — land, fields: St James's street, SPP; le Four Cabot, SA. 'steep' road.

Rue, la **Rouge** — fields, land, road: E of Mont Arrivé to le Bouet, les Vauxlorens, SPP; below les Eturs, C. 'red' road, from reddish gravel.

Rue, la **Verte** — houses, fields: les Quartiers, SSP; les Landes du Marché, VdelE; le Chemin le Roi, SM. 'green' lane.

Rue, la **Vieille** — houses, land: N of les Effards, SSP; in Reservoir, at la Girouette, SSV; N of rue de Laitte, T; la Gachière, SM/SA. 'old' road.

Rue Rocheuse, la — roads: N-E of les Grantez, C; S of les Sablons, SPB. 'rocky' road. Also spelt **Rocqueuse**.

Rue(s), les — land, fields: la Hougue Fouque, les Bordages, SSV; les Caches, SM; les Eperons, SA. 'road(s)'.

Rue(s), les — roads with proper names established over a long period, omitting those 'created' in the 20th c. through parish road signing, are noted as individual entries and listed below.

SPP rue **Allez**, rue **Berthelot**, rue du **Bordage**, rue **du Bouet**, rue **Bunel**, rue **des Camps Collette Nicolle**, rue **de la Chasse Vassal**, rue **des Cornets**, rue **de la Fallaize**, rue **de Fermains**, rue **de la Fontaine**, rue **Forêt**, rue **des Forges**, rue **des Godaines**, rue **de la Grande Marche**, rue **de la Gravée**, rue **de la Guelle**, rue **Havilland**, rue **de la Petite Marche**, rue **du Mau(re)pas**, rue **de Maupertuis**, rue **des Meules**, rue **des Mulettes**, rue **à l'Or**, rue **du Parcq**, rue **Paris**, rue **des Prestres**, rue **du Haut Pavé**, rue **Poidevin**, rue **de la Pouquelaie**, rue **du Pré**, rue **du Puits**, rue **de la Ramée**, rue **des Rocquettes** (**2**), rue **de Rozel**, rue **St Jacques**, rue **St Jean**, petite rue **St Jean**, rue **Tanquerel**, rue **du Truchot**, rue **des Vaches**.

SSP rue **à Chiens**, rue **du Clos**, rue **Colin**, rue **des Dicqs**, rue **du Port**, rue **de l'Eglise**, rue **des Grandmaisons**, rue **du Nocq**, rue **Quartiers**, rue **Roland**, rue **Sauvage**,rue **du Vaugrat**.

CduV rue **Allez**, rue **des Coutances**, rue **Doyle**, rue **Flère**, rue **du Hurel**, rue **Mahaut**, rue **de la Maraive**, rue **du Passeur**, rue **ès Ralles**, rue **Robin**, rue **Qui Tempête**.

VdelE rue **des Annevilles**, rue **des Arguilliers**, rue **des Barrats**, rue **de la Chapelle**, rue **Colin**, rue **du Four**, rue **Maingy**, rue **des Marais**, rue **des Priaulx**, rue **de la Trappe**.

C rue **a l'Angle**, rue **de dessous l'Aumône**, rue **du Barrat**, rue **de la Barsière**, rue **au Beir**, rue **des Bergers**, rue **des Bergerots**, rue **des Besoignes**, rue **de Beuverie**, rue **Boudin**, rue **de la Cache**, rue **des Cailles**, rue **de la Cannevière**, rue **de Carteret**, rue **des Carterets**, rue **de Côbo**, rue **Cohu**, rue **des Covins**, rue **du Crève Cœur**, rue **à l'Eau**, rue **des Effards**, rue **de l'Eglise**, rue

des Emerais, rue **d'Enfer**, rue **Etur**, rue **des Eturs**, rue **au Feyvre**, rue **au Ferré**, rue **du Feugré**, rue **de Français**, rue **du Frocq**, rue **des Genâts**, rue **de la Generotte**, rue **Giffard**, rue **à Gots**, rue **des Grantez**, rue **de la Haye**, rue **des Houmets**, rue **des Hougues**, rue **des Huchettes**, rue **de la Hurette**, rue **des Jaonnets**, rue **Jullienne**, rue **de la Lande**, rue **des Landes**, rue **des Landelles**, rue **de la Mare de Carteret**, rue **des Marottes**, rue **à la Novelle**, rue **des Orais**, rue **Piette**, rue **au Prêtre**, rue **au Querrier**, rue **des Queux**, rue **des Rameurs**, rue **du Rocquer**, rue **à Ronces**, rue **à Saules**, rue **Jean Toulle**, rue **du Tour**, rue **dela Trappe Berger**, rue **des Trenquesous**, rue **des Varendes**, rue **des Vergies**, rue **du Villocq**.

SSV — rue **de l'Arquet**, rue **des Longbos**, rue **des Bordes**, rue **des Longs Camps**, rue **au Cevecq**, rue **des Choffins**, rue **du Clos Landais**, rue **de l'Eglise**, rue **Féveresse**, rue **du Fillage**, rue **ès Frenes**, rue **des Jaignes**, rue **Mahaut**, rue **à l'Or**, rue **du Prier**.

SPB — rue **des Ardaines**, rue **de la Bette**, rue **du Bordage**, rue **de la Boue**, rue **des Brehauts**, rue **du Camp Bailleul**, rue **du Camp Massy Brehaut**, rue **du Camp au Prêtre**, rue **du Campère**, rue **du Câtillon**, rue **du Coudré**, rue **Dan Nicolle**, rue **du Douit Benest**, rue **de l'Eglise**, rue **du L'Eclet**, rue **du Gain**, rue **des Galliennes**, rue **des Godefroys**, rue **du Grais**, rue **des Juliennes**, rue **de Louis**, rue **des Marchez**, rue **de la Claire Mare**, rue **des Marais Gouie**, rue **Meslin**, rue **des Paas**, rue **du Pont**, rue **du Provendel**, rue **de Quanteraine**, rue **des Raies**, rue **ès Sarreries**, rue **du Vallet**.

T — rue **du Banquet**, rue **de Beuval**, rue **des Buttes**, rue **du Camp d'Ebat**, rue **Camp Henry**, rue **Clos aux Pommiers**, rue **de Courtil Laurent (Laurens)**, rue **des Coutures**, rue **de l'Eglise**, rue **de la Folie**,

	rue **de dessous les Hures**.
F	rue **des Auberts**, rue **du Bordage**, rue **Bouillonnière**, rue **des Grons**, rue **des Landes**, rue **des Reines**.
SM	rue **du Bordage**, rue **de Calais**, rue **Cammus**, rue **des Camps**, rue **Cauchez**, rue **de l'Eglise**, rue **des Escaliers**, rue **Fainel**, rue **de la Fallaize**, rue **au Gouais**, rue **des Grons**, rue **du Hurel**, rue **des Landes**,rue **des Maindonnaux**, rue **de la Marette**, rue **du Navet**, rue **des Orgeris**, rue **de la Personnerie**, rue **Picart**, rue **de Bon Port**, rue **Poudreuse**, rue **de la Quevillette**, rue **de Saint**, rue **de la Vallée**, rue **des Varinvarays**, rue **de la Ville Amphrey**.
SA	rue **des Blicqs**, rue **des Bémonts**, rue **de la Croix au Baillif**, rue **de l'Echelle**, rue **des Fourques**, rue **Marquand**, rue **des Morts**, rue **des Pointes**.
Rue, de **Lucas de la**	field: le Basses Capelles, SSP. Extant surname **De La Rue**.
Rue, de **Nicolas de la**	field: les Rouvets, VdelE. Derivation as above.
Ruel, le	land: le Bouet, les Terres, N of Fermain, rue de la Fontaine, Glategny, SPP; la Rousse Mare, SPB; Icart, SM. 'minor stream or runnel'. douit, ruisseau, russel q.v.
Ruelle(**s**), le(s)	fields: rue des Bergers, la Pitouillette, C. former 'water lane'.
Ruette(**s**), la(les)	houses, land, roads, all parishes. 'lane'.
Ruette(**s**)	with proper names established over a long period, omitting those created in the 20th c. through parish road signing, are noted as individual entries and listed below.
SPP	ruette **Bataille**, ruette **des Belles Filles**, ruette **de la Bouillonne**, Ruette **Fleurie Hailla**, ruette **de la Fontaine des Corbins**, ruette **de Havelet**, ruette **des Maisons**

	Brulées, ruette **Meurtrière**, ruette **du Parcq**, ruette **des Rocquettes**, ruette **des Venelles**, ruette **desVauxlorans**.
SSP	none.
CduV	ruette **de Camp Dry**, ruette **des Corvées**.
VdelE	none
C	ruette **des Cherfs**, ruette **du Frocq**, ruette **des Houguettes**, ruette **du Moulin des Niaux**, ruette **de St George**, ruette **des Vallées**, ruette **du Villocq**.
SSV	ruette **du Mont Varouf**, ruette **de la Potère**.
SPB	ruette **Thomas Henry**, ruette **du Galley**, ruette **des Hougue Falle**, ruette **des Paas**, ruette **Suart**, ruette **du Vallet**, ruette **Vasse**.
T	ruette **de l'Eglise**.
F	ruette **du Gal**, ruette **de la Roberge**.
SM	ruette **Bailleul**, ruette **à la Bellette**, ruette **du Fainel**, ruette **sur Fermains**, ruette **des Frènes**, ruette **du Navet**, ruette **Rabey**.
SA	ruette **de l'Ecole**.
Ruettes Brayes	see **Brayes**, ruette.
Ruine, la	house: rue de l'Eglise, C. 'in decay'.
Ruisseau, le	fields: les Rébouquets, la Villette, la Ville Amphrey, SM. 'stream'. douit, ruel, russel q.v.
Rulées, les **Camps**	field: les Laurens, T. 'marked' strips of land.
Ruley, le **Camp**	field: le Niaux, C. Derivation as above.
Rulley, le	field: les Videclins, C. Derivation as above.
Rungefer, de	house, land: S of rue du Clos, SSP; rue du Passeur, CduV. Surname extinct 15th c. **Rungefe**r. Mis-spelt **Rongefer**, **Roncefer**. Also feudal holding of land, Bordage **Rungefer**, E of la Ramée, SPP.
Rurée, de	land: Pleinmont, T. Origin unknown.

Russel, le — house, fields: N of le Chene, F; le Creux Mahie, F; la Villette, Sausmarez Manor, SM. 'stream'. Also spelt **Ruissel**, **Ruizel**, **Ruzel**. douit, ruel, ruisseau q.v.

S

Sablon(**s**), le(les) — houses, land: la Moye, les Grippios, CduV; S of l'Erée headland, SPB; S of les Houards, F. 'fine sand'.

Sablon, **Fosse** à — land: near le Préel, C.

Sablonnière, la — house, land: la Mare de Carteret, C. 'sand pit'.

Saette, le **Bordage** — site: unlocated, possibly E of les Huriaux, SM. Held by **Matthieu Saette**, 1331. Surname extinct 14th c. **Saette**. bordage q.v.

Saffran, à — land: Bordeaux, CduV; la Maison de bas, VdelE; les Hougues, C; la Maison de haut, SPB; le Hurel, la Fosse, SM. 'saffron'. Introduced in 14th c, and cropped until 16th c.

Sage, la **Croix** au — field: les Sages, T. croix q.v.

Sage, la **Croûte** au — houses: land, North side, CduV. croûte q.v.

Sage, la **Hougue** au — field: N of la Rochelle, CduV. hougue q.v

Sages, les — houses, fields: les Sages, SPB/T. Surname extinct 17th c. **Le Sage** (extant Jersey).

Saille, la — land: S of le Murier, SSP. Probably seille, i.e. seigle, 'rye'.

Saint(**s**), de — large district: Saints, SM. Surname extinct 14th c. **Sayn**. Also spelt **Sainc**, **Saing**, **Saynk**, **Sayns**.

Saint(**s**), **l'Eperquerie** de — site: area of harbour, SM. eperquerie q.v.

Saints, rue de — road: S of rue des Escaliers, SM.

Saints, — listed below are references to buildings and land, excluding the parish churches which are noted under Eglises.

St André, de —
1. field: S of the 'Track', SSP. Owed feudal dues to Fief le Roi, SA.
2. fields: le Tertre, Ste Hélène, SA. Presumably owed rente or dîme (tithe) to church or living of the parish.

Ste Anne, de —
1. houses, land: la Porte, C. Formerly part of the **Manoir de Ste**. **Anne** q.v.
2. fields: la Lande, N of l'Aumone, C. Either formerly part of above **Manoir de Ste Anne** or owed rentes to a church guild.

Ste Appoline, la **Chapelle** de — chapel: la Grande Rue, SSV. The only medieval chapel to have survived intact. Built about 1392 by Nicolas Henry, of le **Manoir de la Perelle**. Chapelle, manoir de la Perelle q.v.

St Briocq, Croute — land: rue des Galliennes, SPB.

St Brioc(q), la **Chapelle** de — site: St Briocq, SPB. chapelle, croûte q.v.

St Briocq, la **Hure** de — land: St Briocq, SPB.

Ste Catherine, de —
1. house, land: la Ramée, SPP. 18th c farm house.
2. land: les Petils, CduV. In name of Ste Cathelinne, 1591. Presumably owed rente to a church guild.

St Clair, **Fontaine** — spring: St Clair, SSP.

St Clair, la **Chapelle** de — site: St Clair, SSP. Remains presumably incorporated in present house.

St George, de — fields: E of St George, C.

St George, Fontaine — spring: St George, C.

St George, la **Chapelle** de — site: St George, C. Footings and parts of wall remain in garden. Earliest record dates from 1156. Used as parish school from

	1675 until in demolished 1769. Also feudal holding of land, Fief de la **Chapelle Saint Georges**, C. with feudal dues originally payable to the chaplain. chapelle q.v.
St George, **Ruette** de	land: la Rocque ès Boeufs, C.
St George's Esplanade	road: W of la Salerie, SPP. 19th c. widening of coast road.
St Germain, **Croix**	cross: near le Tertre, C.
St Germain, la **Chapelle** de	site: W of Castel school, C. Totally quarried away in 19th & 20th centuries. chapelle q.v.
Ste Hélène, de	house, land: W of St Andrew's church, SA. Name taken from fief held by the priory of 'Ste Hélène in Hagua', le Cotentin, France. (Fief Ste Hélène).
Ste Hélène, Fief de	see **Ste Hélène**.
St Jacques, de	land: la Ramée, SPP; le Douvre, T. Presumably owing rentes to either la chapelle St Jacques or a church guild.
St Jacques, la **Chapelle** de	site: near Mon Plaisir, SPP. Name presumably commemorated pilgrimage to St James of Compostella, Spain. chapelle q.v.
St Jacques, la **Pièce**	land: S of le Douvre, T. Presumably owed rente to a church guild.
St Jacques, rue de	road: W of Elm Grove, SPP.
St Jean, de	land: la Haye du Puits, C. Presumably owed rentes to a church guild.
St Jean du Bosq, la **Chapelle** de	site: le Bosq, SPP. Site lost. chapelle q.v.
St Jean de la Houguette, la **Chapelle** de	site: now occupied by St Martin's school, SM. chapelle q.v.
St Jean, **Petit** rue de	road: S of Union Street, SPP.
St Jean, rue de	road: S of the Grange, SPP.

St Julien, la **Chapelle** de	site: at bottom of St Julian's avenue, SPP. chapelle, école, la petite q.v.
St Magloire, la **Chapelle** de	site: S of les Beaucettes, CduV. Recorded in 1156, became **St Mallière** in local record, 1591, 1654, and **St Magloire** from 1727 onwards. chapelle q.v.
Ste Marie du Marais la **Chapelle** de	site: le **Château des Marais**, SPP. chapelle q.v.
St Martin, **Franc Fief**	feudal holding of land. SPP / SM.
St Martin, **Fontaine**	spring: Jerbourg, SM.
St Michel, le **Pont**	crossing: main crossing of le Braye du Valle, west of Vale church. See also Pont Allaire for main crossing to east of that church.
St Michel, Fief de	feudal holding of land. Originally only le Clos du Valle but now also in VdelE; C; SSV; SPB and T.
St Nicolas, de	see **Senicolas**.
St Pierre, la **Croix**	site: le Long Frie, SPB. Wayside cross on corner of road facing down to the church, opposite feudal court seat of Fief Suart. croix q.v.
St Pierre, la **Maison**	site: les Reveaux, SPB. House demolished & replaced 19th c.
St Peter's valley	houses, land: les Ruettes Brayes, SM. 19th c.
St Sauveur, la **Maison**	site: les Reveaux, SSV. House demolished 19th c., site remains a field.
St Thomas d'**Anneville**	site: les Annevilles, SSP. Site lost.
Saisine, la	field: l'Alleur, T. 'seisin', i.e. taking a right of legal possession on non-payment of rentes.
Salerie, la	site: N of Glategny, SPP. 'salting place for fish'. Also fishing harbour and fort. Fort q.v.
Saline(s), la(les)	land, fields: rue du Clos, W of St Clair, SSP; la Mare de Carteret, C; la Mare Hailla, SPB.

	'salt pans', late medieval. house: house at le Coin Fallaize, SM. 19th c. name.
Salle, la	fields: le Valinguet, C; le Douit, la Clôture, SPB. Surname extinct 14th c. **De La Salle**.
Salle, le **Clos** de	former field: S of la Mare de Carteret, C. Derivation as above. clos q.v.
Sallemon, de	fields: rue à l'Or, SPP; les Salines, SSP; les Arguilliers, VdelE; les Cambrées, le Coudré, SPB. Extant surname (abbreviated) **Salmon**. ville q.v.
Sallemon, le **Bordage**	feudal holding of land, unlocated, CduV. Bordage q.v.
Salliot, la **Mare**	land: la Croûte au Sage, CduV. Surname extinct 15th c. **Salliot**.
Salt Pans	1. land: W of la Grosse Hougue, SSP. Area set inside former Braye du Valle (Wm. Gardner's Survey 1787) connected by a canal to the harbour for sea water, used 1806–40s, then abandoned. Nocq q.v. 2. land: W of le Grand Fort. Formerly late medieval salt pans, now a housing development.
Samarez, de	see **Sausmarez**.
Sambue, la	land: Pleinmont, T. Origin unknown.
Sampson, le **Camps**	land: near la Fontaine, SSV.
Samson, de	field: les Trachieries, SSP. Surname extinct 16th c. **Samson**.
Samson, de **Pierre**	house, field: rue de Bouverie, C. Derivation as above.
Samson, la **Hougue**	land: near Baugy, CduV. Derivation as above.
Samson, le **Clos**	fields: la Fontaine, SSV. Derivation as above.
Samson, les **Camps**	fields: la Fontaine, SSV. Derivation as above.

Samsonnet, de	land: les Grandmaisons, SSP; Bordeaux, CduV. Surname extinct 16th c. **Samsonnet**. Also spelt **Chansonnet**, **Samsonnés**.
Samsonnet, la **Hougue**	land: S of harbour, SSP. Derivation as above. hougue q.v.
Samuel, de	land: le Villocq, C. First recorded 1853. Forename (m) **Samuel**.
Sandre, de **Pierre**	field: le Varclin, SM. Surname extinct 18th c. **Sandre**.
Sandre, la **Croûte**	house, land: rue Poudreuse, SM. Derivation as above.
Sarel school	building: la Grande Rue, SSV. After Major General H A Sarel, Lieutenant Governor, 1883–1885. Presently St Saviour's parish Douzaine room.
Sarre, de **Pierre**	field: la Neuve Maison, SPB. Extant surname **Sarre**.
Sarreries, les	fields: above la Neuve Maison, SPB. Former house replaced by la Neuve Maison. Derivation as above, in plural form.
Sarreries, rue des	road: unlocated, SPB.
Sauce, le **Mont**	fields: W of Icart, SM. Surname extinct 15th c. **Sause**.
Sauche, à	see **Sauge**, à.
Sauchet, le	house, fields: le Pont, T; Jerbourg, SM. 'shallow-ploughed'. Also spelt **Sochel**, **Sochet**.
Saudrée, la	land, all parishes. 'willow bed'. Also spelt **saudraye**, **saudreye**, **sauldraye**, **sauldrée**, **sodrée**.

Saudrées de **Janin Bernard**, les	les Marais, SSP.	Bernard q.v.
Saudrée des **Cherfs**, la	les Cherfs, C.	cherfs q.v.

Saudrée des **Mourières** les	les Rohais, SPP.	mourières q.v.
Saudrée Ecollasse, la	les Baissières, SA.	ecollasse q.v.
Saudrée ès **Mottes**, la	E of la Ramée, SPP.	mottes, q.v.
Saudrées du **Bouillon** les	les Mourains, C	bouillon q.v.

Sauge, à — land: le Clos Hoguet, SSV. 'sage'.

Saules, rue des — road: unlocated, C.

Saumarez, de — houses, fields: Saumarez park, C.

Saumarez, le **Fort** — fortification, martello tower: l'Erée, SPB. Built early 19ttc. Extant surname **Saumarez**. Formerly known as L'Erée tower. Named in late 19th c after Admiral Lord de Saumarez. Also feudal holding of land, **Fief de Saumarez**, C. Fort, Martello tower q.v.

Saunier, au — field: les Beaucamps, C. Surname extinct 15th c. **Le Saulnier**.

Saunier, le **Camp** au — land: Icart, SM. Derivation as above. camp q.v. Also feudal holding of land, Fief au **Saunier**, C.

Sausmarez, de — houses, fields: Fort George, SPP; le Rocher, VdelE; rue à Ronces, C; Sausmarez manor, SM; les Eperons, SA. Extant surname **De Sausmarez**. Also feudal holding of land, Fief de **Sausmarez**, SM.

Sauvage, la rue — road: N of les Capelles, SSP. On W side of late medieval salt pans (les Salines). 'wild, rough road'.

Sauvagées, les — houses, land: l'Islet, E of Baubigny, SSP. Surname extinct 17th c. **Sauvaget**, **Chauvaget**.

Sauvarin, de — house, land: les Salines, SSP. Extant surname **Sauvarin**.

Sauvarinerie, la — house: les Grands Moulins, C. Derivation as above.

Sauvary, de	field: le Marais, Cdu V. Extant surname **Sauvary**.
Saux, à	land, fields: le Déhus, CduV; les Arguilliers, VdelE; le Variouf, F. 'willow'.
Saulx, la **Croûte** à	field: Maison de bas, VdelE. Derivation as above. croûte q.v.
Saward school	building: les Bordages, SSV. After Major General M H Saward, Lieutenant Governor 1899-1903. Now a private house.
Sayve, la **Ville**	see **Ville Sayve**, la.
Scala, le **Moulin**	see **Echelle**, le **Moulin** de l'.
Scevallet, le	see **Cevallet**, le.
Sebire(**s**), de(s)	fields: rue d'Aval, S of les Maingys, VdelE. Extant surname **Sebire**.
Sechage(**s**), le(s)	land: low lying coastal areas, drying places for vraic, including burning.
Secheur(**s**), le(s)	alternative expression for **sechage**.
Seigle, à	fields: la Varde, SPP; les Gigands, SSP; Portelet, T. 'rye'. Also spelt **seille**.
Seigneurie, la	1. land: les Vallettes, rue à l'Or, SSV. Probably owed a rente to seigneur of fief. 2. house: Pleinmont, T. Name changed in mid-19th c. by Pierre Robilliard, seigneur of four fiefs.
Séjour, Bas	house: les Fries, C. 19th c. villa.
Séjour, Beau	former house: la Butte, SPP, demolished and replaced with leisure centre in 1970s; la Dévise, SPB. 19th c. villas.
Séjour, Haut	house: les Hougues, C. Small 19th c. villa.
Sellery, le	field: les Martins, SM. Origin unknown. Possibly from **Essilleril**, **Esillery**. Silleries q.v.
Selline, de	field: le Marais, CduV. Surname extinct 16th

	c. **Asseline**. Recorded **Asseline**, 1654. Misspelt since 18th c.
Semyonne, la	see **Simeonne**.
Senay, de	field: rue à l' Or, SPP. Surname extinct 16th c. **Seney**.
Sénéquins, les	fields: N of l'Aumône, C. Origin unknown.
Senicolas, de	field: la Houguette au Comte, C. St Nicollas. Presumably from a rente owed to a church confraternity. Misspelt since 18th c.
Sensière, la	see **Censière**.
Sept Rocques, le **Clos** des	see **Rocques**, le **Clos des Sept**.
Serceuil, le	see **Serqueur**, le.
Sergentée, la	house, land: E of le Villocq, C. Past association with 'sergents' of Fief St Michel, C.
Sermonier, la **Vallée** au	land: on SW side of headland at l'Erée, SPB. Origin unknown. Earliest record 1504. rocque, vallee q.v
Serqueur, le	land: les Houmets, C. Origin unknown. Spelt as **Serqueur** 16th & 17th centuries, misspelt as **Cerceuil**, **Serceuil** since 18th c.
Serquier, **Camp** du	land: rue des Emrais, C.
Servais Massy, de	see **Massy**, de **Gervais**.
Servée, de	land: les Queux. C. Surname extinct 16th c. **Gervaise**. Spelt **Gervays** to 1634, misspelt **Servais**, **Servée** since 1662.
Seterie, la	land: W of les Abreuveurs, SSP. Origin unknown.
Seulx, les	field: les Vesquesses, C. 'elder tree'.
Sevallet, le	see **Cevallet**, le.
Sevée, sur la,	land, field: le Bourg, F; Icart, SM. Surname extinct 13th c. **Sayve**, **Seyve**. Also spelt **Sevet**. Ville Sayve q.v.

Silleries, les — field: le Chêne, F. possibly 'deep furrow'. **Essilleril**, **Esillery**. Sellery q.v.

Sillon, le **Long** — land: les Fries, C. 'long furrow'.

Simeonne, la — field: les Jetteries, SSV. Probably forename (f) **Simone**. Recorded **Semyonne**, 1604.

Simon, de — fields: la Grande Marche, le Vauquiédor, SPP; les Trachieries, SSP; les Vauxbelets, SA. Extant surname **Simon**.

Simon, de **Jean** — field: W of la Gron, SSV. Derivation as above.

Simon, la **Vallée** — field: les Clercs, SPB. Derivation as above. vallée q.v.

Simon, le **Clos** — land: S of la Mare de Carteret, C. Derivation as above. clos q.v.

Simonne, de — fields: les Varendes, C; le Becquet, SM. Derivation as above.

Simons, les — house: les Simons, N of la Croisée, T. After **Jean Simon**, 1597.

Sion — apartments: les Brehauts, SPB. Former Methodist chapel.

Smith street — see **Forges**, la **rue** des.

Sochel, le — see **Sauchet**, le.

Sochet, le — see **Sauchet**, le.

Sohier, de — house, fields: les Trachieries, SSP; S of Baugy, CduV; les Delisles, C; les Reveaux, SPB; les Bourgs, SA. Surname extinct 19th c. **Sohier** (extant Jersey).

Sohier, de **Robin** — field: W of les Rohais de haut, SA.

Solier(**s**), de — land, fields: la Folie, le Marais, CduV; Pleinmont, T. Origin unknown.

Soif, **Port** — see **Port Soif**.

Solidor, de — house, fields: la Folie, CduV. Origin unknown. Spelt **Sollidor** 1591.

Sommeilleuses, les	cliff land: W of Petit Bot, F. 'dormant land'.
Sommet de la **Petite Porte**, **Camps** au	see Porte, camps au Sommet de la Petite. SM.
Sommet, le	land: le Petit Port, Jerbourg, SM; les Monts Héraults, SPB. 'summit'.
Son, la **Hougue** au	hillock: W (above) les Canichers, SPP. Probably a sounding hillock, see la Rocque qui sonne. hougue q.v.
Sottevast	see **Sotuas**.
Sotuas, la **Hougue**	land: la Hougue du Pommier, C. Surname extinct 14th c. **Sottevast**. hougue q.v. Also feudal holding of land, Fief de **Sotuas**, C.
Soucique, la	house, field: W of le Chêne, F. 'marigold'.
Sounil, de	see **Founin**.
Source, la	field: rue du Tour, C. Recorded 1634. 'source' of spring.
Souscription, **Fort**	site: les Hougues, E of rue Paris, & N of rue du Puits, SPP. 18th c. fort, recorded Fief le Roi SPP 1837. Demolished late 19th c. and quarried for stone. Interior currently used as paved car park.
Sous l'**Aumône**	see **Aumône**, l'.
Sous l'**Eglise**	see **Eglise**, **Sous** l'.
Sous les **Courtils**	see **Courtils**, **Sous** les.
Sous les **Hougues**	see **Hougues**, **Sous** les.
Spignolet, de	field: la Bertozerie, SPP. Origin unknown.
Springfield	house: the Queen's road, SPP. Built by Thomas Gosselin, early 19th c.
Star Fort	fort: W end of Pembroke Bay, CduV.
Sto', de	fields: Fort George, SPP; les Salines, SSP; la Croix au Baillif, SA. Extinct surname **Setost**. Spelt variously **Setost**, **Stoc**, **Stoo**, **Stoon**, **Stost**, **Stove**, **Stow**.

Strand, the — lane: above South Esplanade, SPP.

Suart, de — fields: la Pomare, les Paysans, SPB. After Fief Suart, on which fields are situated.

Suart, Fief de — feudal holding of land, SSV / SPB.

Suart, la **ruette** — lane: les Paysans, SPB. On feudal holding of land, Fief Suart, SSV/SPB, after **Roger Suart**, late12th c.

Suart, les **Camps** — fields: rue de l'Arquet, SPB. In full, "**les Camps du Fief Suart**", i.e. strips of land on Fief Suart.

Suchez, la **Hougue** à — land: unlocated, CduV.

Sud, du — fields: S of les Cotils, SPP; S of le Val, SPB. Literally 'to south of'.

Surelles, les — field: le Pont Renier, SPB. 'sheep's sorrel'.

Surre, la — fields: les Raies, SPB; unlocated, SM. After **Martin le Sur**, living in SPB, 1551.

Susanne, de — fields: le Moulin de haut, C; les Blicqs, les Gouies, SA. Forename (f) **Suzanne**.

T

Tablet, la — house, fields: la Lande, C; E of les Clercs, SPB; Bon Port, SM. 'level area of land', in a valley. Also spelt **Tabler**, **Tablier**.

Tablet, la **Vallée** du — land: le Tablet, SPB. Derivation as above.

Tablette, la — land: Perelle Coast road, SSV. 'small level area of land'. Area adjacent to beach, for boats and drying of vraic. Now built on, with two dwellings.

Taive, la **Hougue** — land: S-W of Pleinmont, T.

Talbots, les — house, land: N-W below les Niaux, C. Surname extinct 15th c. **Tallebot**. Name given in 19th /20th centuries to whole valley from Les Poidevins to les Grands Moulins.

Tallut, le	land, fields: les Fauquets, la Bernauderie, C; Sous l' Eglise, SSv. 'piece of raised ground, bank'. Also spelt **Talum**, **Talus**.
Taniole, la	see **Tonniole**, la.
Tannerie, la	land: le Pont Renier, SPP; les Tilleuls, VdelE. House now demolished. 'tannery'.
Tanquerel, de	land, street: les Guelles, Church square, SPP. Extant surname **Tanquerel**.
Tardif, de	fields: la Quevillette, les Caches, SM. Extant surname **Tardif**.
Tardif, de **Martin**	field: Sausmarez manor, SM. Derivation as above.
Tardif, le **Douit** ès	stream: W of Bon Port, SM. douit q.v.
Tardif, le **Mont**	land: les Nicolles, SM. Derivation as above. mont q.v.
Tasseine, de	field: l'Aumône, C. 'heaping'.
Tassières, les	fields: la Villiaze, F. Derivation as above.
Taudevins, les	house, land: les Landes, SPB. Recorded les **Tostevins**, 1807, 1854, les **Tautevins**, **Taudevins** since 1873. Extant surname **Tostevin**.
Taviole, la	see **Faviole**, la.
Tée, la	see **Thée**, la.
Téhy, le **Clos**	see **Théhy**, le **Clos**.
Tellier, la **Houguette** au	land: le Tertre, C. Surname extinct 16th c. **Le Tellier**. Also spelt **Le Tillier**. houguette q.v. Also feudal holding of land, Bouvée des **Telliers**, C.
Tempête, la **rue qui**	road: Bordeaux, CduV. 'bumpy road'.
Temple, le	'temple': description used in 16th, 17th & 18th centuries for all parish churches, as église, q.v. was considered to be 'papist'.
Tenaisie, la	land: les Blanches, SM. 'tansy'.

Terre à Bœuf, la	land: les Maindonnaux, SM	Beuf q.v.
Terre au Guillard, la	land: les Baissières, SPP	Guillard q.v.
Terre au Mière, la	land: les Rohais, SPP	Mière q.v.
Terre au Querpentier, la	land: les Rohais, SPP	Querpentier q.v.
Terre Dolbel, la	land: E of la Ramée, SPP	Dolbel q.v.
Terre Douglas, la	land : les Rohais, SPP	Douglas q.v.
Terre Fouachin, la	land: la Couture, SPP	Fouaschin q.v.
Terre Guillemet, la	land: unlocated, F	Guillemet q.v.
Terre Hugues, la	land: la Couture, SPP	Hugues q.v.
Terre Lesant, la	land: la Couture, SPP	Lesant q.v.
Terre Norgeot, la	land: E of la Grande Rue, SSV	Norgeot q.v.
Terre Perquin, la	land: l'Hyvreuse, SPP	Perquin, q.v.
Terre Sauvaget, la	land: la Vrangue, SPP	Sauvagées q.v.
Terre Thomas Cotil la	land: les Terres, SPP	Costil q.v.

Terre, la **Basse** — field: les Sages, T. 'bottom land'.

Terres Gosselins, les — land: les Terres, SPP Gosselin q.v.

Terres Neuves, les — land: les Etibots, les Canichers, Hauteville, SPP. 'new' land, i.e. reclaimed into use.

Terres, les **Blanches** — fields: E of la Claire Mare, SSV. 'white land, clay'.

Terres, les **Franches** — land: l'Echelle, SA. 'land free of feudal dues' on Fief Ste Hélène, q.v.

Terres, les — land: Belvedere house Fort George, SPP. 'lands' in strips, pieces, fields, i.e. land not fully enclosed. Val des Terres q.v.

Terres, les **Val** des — thoroughfare: les Terres, SPP.

Territoire, le — locality, island wide. Land, fields, in same area. I.e. in a 'contrée', q.v.

Tertre, le	houses, land, fields: S of la Croix au Bois, CduV; le Hamel, VdelE; N of Castel school, la Rivière, les Videclins, la Trappe Berger, les Grantez, C; le Mont Varouf, SSV; S of les Bourgs, SA. 'knoll' on a hillock or intermediate ground. hougue q.v.
Testard, la **Hougue**	land: les Monts (Delancey park), SSP. Surname extinct 14th c. **Testard**. bordage, hougue q.v. Also feudal holding of land, Bordage **Jourdain Testard**, SSP.
Tête, la	field: la Houguet (les Prevosts), SSV. At 'head' of.
Thée, la	land: S of Victoria avenue, SSP.
Thée Mauger, la	field: part of la Thée, above. SSP. Crown fee farm, formerly part of le Marais d'Orgeuil Fee farm. marais, Mauger q.v.
Théhy, le **Clos**	field: opposite Seaview farm, C. Surname extinct 14th c. **Théy**, **Théhy**. clos q.v.
Thomas Brehaut, de	see **Brehaut**, de **Thomas**.
Thomas Breton, de	see **Breton**, de **Thomas**.
Thomas, de	field: la Vrangue, SPP. Either forename (m) or surname **Thomas**.
Thomas, de **Jean**	field: N of les Emerais, C. Surname **Thomas**. Also, by repute, dune & former slipway at centre of Vazon bay, C.
Thomas de la Court, de	see **Cour**, de **Thomas De La**.
Thomas de la Landelle, de	see **Landelle**, de **Thomas De La**.
Thomas Henry, de	see **Henry**, de **Thomas**.
Thomas Henry, la **ruette**	lane: les Adams, SPB.
Thomas Néel (**Nez**), **Croûte**	land: rue des Camps, SM.
Thomas Néel (**Nez**), de	see **Néel** (Nez), de **Thomas**.

Thomas Le Ray, de — see **Ray**, de **Thomas Le**.

Thomas Queripel, de — see **Queripel**, de **Thomas**.

Thomasse, de — field: Bon Port, SM. Forename or surname **Thomasse**.

Thomasse, de **Pierre** — fields: le Douit du Moulin, les Marchez, SPB. Surname extinct 16th c. **Thomasse**.

Thomasse De Lisle, de — see **Lisle**, de **Thomasse De**.

Thomassin, de — see **Toumassi**, le.

Thomellin, la **Vallée** — field: les Hèches (la Gallie), SPB. Forename **Thomellin** pet form of Thomas.

Thoume, de — former field: les Granges, SPP. Extant surname **Thoume**. Also spelt **Thommes**, **Thoumes**, **Tommes**, **Tomez**, derived from forename (m) **Thomas**. **Thoume**, de **Catherine** field: le Chêne, F.

Thoumelin, de — fields: les Cambrées, T; les Reines, F. Thomellin q.v.

Thoumine, de — field: les Huriaux, SA. Extant surname **Thoumine**, from dimimutive of Thomas.

Thybai, le — land: la Mouranderie, T. Origin unknown.

Tiault, de — fields: unlocated, SPP; la Houguette, C; la Bellée, T. Surname extinct 16th c. **Tyault**. Also mis-spelt **Tiost**. Tiosterie q.v.

Tielles, les — fields, cliff land: les Tielles, T. Surname extinct 15th c. **Tielle**.

Tiété, le — flelds: le Frie, SPB. Surname extinct 16th c. **Le Thetie** (**Thomas Le Thetie**, 1470). Also spelt **Le Thietey**.

Tiffaigne, le **Clos** — field: W of les Maumarquis, SA. Also spelt **Tiphaigne**. Misspelt **Fiffaigne**, 19th & 20th centuries. Derivation as above.

Tiffaignes, les — fields: N of la Lande, C; S of les Prevosts, SSV. Surname extinct 15th c. **Tiphane**, **Typhane**. After **Richard Tiphane**, jurat, 1350–1367.

Tilleuls, les	houses: les Landes du Marche, VdelE; les Merriennes, SM. 'the limes'. Named in late 19th c.
Tillier, la **Houguette** au	see **Tellier**, la **Houguette** au.
Tintellée, la	fields: N of rue des Pointes, SA. Possibly from 'tenter', an area where cloth was stretched after fulling.
Tiosterie, la	house: E of la Porte, C. From the surname **Tiost**, a form of **Tiault**, q.v. Formerly called Beaumont.
Titus, de	field: les Sauvagées, SSP. Forename (m) **Titus**.
Tolle, la **Ville**	see **Villetole**.
Toll(e)y, de	see **Joli**.
Tombe, la	field: le Coin Fallaize, SM. 'tomb, grave'.
Tonnelle, **Douit** de la	stream: continuation of the Douit des Orgeris to the Braye du Valle, SSP. Orgeris q.v.
Tonnelle, la	land: lower Ruettes Brayes, SPP; les Banques, SSP; les Grands Moulins, C; le Bas Marais, SSV; les Maindonnaux, Bon Port, SM. stone 'tunnel' for a stream. douit q.v.
Tonnelerie, la	see **Etonnelerie**, **Lestournelerie**.
Tonniolle, la	fields: les Paysans, SPB; Farras, F. 'cooperage'.
Torcamp, le	see **Camps**, les **Tors**.
Tor Camp, la **Houguette** du	land: le Dehus, CduV.
Torode, de	field: les Sages, SPB. Extant surname **Torode**. Spelt **Thouraulde**, **Touraude**, 15th & 16th centuries.
Torquetil, de	land: les Sablons, SPB. Surname extinct 15th c. **Torquetil**. Also feudal holding of land, Bouvée **Torquetil**, SPB/T.

Tors Camps, les — see **Camps**, les **Tors**.

Tors Fosse, le — see **Fosse**, le **Tors**.

Torte, à la — field: la Bellieuse, SM. 'twisted, crooked'.

Torte Pièce, la — see **Pièce**, la **Torte**.

Torteval, de — parish: 'twisted valley', from the latin form.

Tortuée, la — field: l'Hyvreuse, SPP. Derivation as above.

Torval, le — house, land: Talbot valley, C. 'twisted valley'.

Tostein, de — see **Toustain**.

Tostevins, les — see **Taudevins**, les.

Totenez, les **Rocques** de — house, field: les Rocques, T. 'rocks capped'. Earliest reference **Roques totannées**, 1715.

Touillets, les — house (grande maison), fields: S of le Préel, C. Surname extinct 16th c. **Toulle**, **Toulley**.

Toulle, la **Croix Jean** — land: S of 'Fairfield', (les Touillets), C. Derivation as above. croix q.v.

Toulle, la rue **Jean** — road: E of les Delisles, C. Derivation as above. rue q.v. Also feudal holding of land, Bouvée **Jean Toulle**, C.

Toulle, le **Camp** — land: les Terres, SPP. Derivation as above. camp q.v.

Toumassi, le — field: W of le Camp Tréhard, SA. Forename (m) **Thomassin**.

Toumelin, de — see **Thoumelin**.

Toumes, de — see **Thoume**.

Tour, la —

1. site of tower: top of rue des Cornets, SPP, la Tour de Beau regard. Beauregard q.v.
2. site of tower: lower le Pollet, SPP, la Tour Gand. Gand q.v.
3. towers: Martello towers, Pre-Martello towers, Victoria tower, q.v.

Tour, le — field: les Blanches Rocques, C. 'sharp turn, bend', in the road.

Tour, ruette de la — road: E of la Bouverie, C.

Tourailles, les — fields: E of les Monts (Delancey park), SSP; W of les Picques, SSV; le Coudré, SPB; le Bigard, F; Saints, SM. 'turns, bends' on the contours.

Tourdin, de — fields: la Vrangue, SPP; les Français, CduV. Presumably from an extinct surname.

Tourelle, la —
1. house: E of les Brehauts, SPB. Recorded la **Tourelle** 1742. 'turret'.
2. house: Saints, SM. Name derived from **Touraille**, 1666.

Tourgand, la — see **Gand**, **Tour**.

Tourgie, la — see **Tourgisse**, la.

Tourgis, de — land, fields: la Vrangue, les Vardes, SPP; les Comtes, la Neuve Maison, SSV; le Variouf, F. Surname extinct 19th c. **Tourgis**.

Tourgis, **Fontaine** — spring: N of la Vrangue, SPP.

Tourgis, la **Hougue** — land: near les Eperons, SA.

Tourgis, le **Frie** — land: le Préel, C. Spelt **Tourgy**, 19th & 20th centuries.

Tourgisse, la — house, land: E of le Bourg, F. Extinct surname **Tourgis** feminised. Spelt **Tourgie** since early 20th c.

Tournel, le — see **Etournel**, l'.

Tourteau D'Eparty, au — field: les Monts Heraults, SPB. The 'Tourteau bequest', presumably donation of a rente or piece of land to the church.

Tourteau Departy, fosse — land: le Mont Herault, SPB.

Tourtel, de — fields: les Blanches Pierres, les Caches, la Hougue, la Marette, la rue au Cammus, SM. Extant surname **Tourtel**.

Toustains, les — fields: les Frances, SSV. Surname extinct 14th c. **Tostain**, **Toustain**. Also feudal holdings of land, Bouvées **Jean** and **Richard Toustain**, SSV.

Touzées, les — see **Déhuzet**, le.

Trac, le **Long** — fields: W of la Marette, SM. 'long tract' of land, not a way or thoroughfare.

Trachieries, les — district: W of l'Islet, SSP. After **Collas** and **Pierre Trachy** 1700. The first 'i' often omitted in present day spelling.

Trachy, de — field: S of Mont Cuet (Cuhouel), CduV. Surname extinct 20th c. **Trachy**.

Trachys, la **Hougue** des — land: S of les Effards, (Anneville), SSP. Derivation as above.

Traculet, le — land: rue à l'Or, les Etibots, les Baissières, SPP; Beauvalet, SSV; Courtil Guilbert, SPB; la Mouranderie, les Simons, T. Diminutive of trac, 'a sliver of land'.

Tranquesous — see **Trenquesous**.

Trappe au **Prevost**, la — field: les Rocquettes, SSV. Prevost, le, q.v.

Trappe Berger, la — house, land: les Generottes, C. After **Jehannet Berger**, 1470. berger q.v.

Trappe Berger, rue du — road: unlocated, C.

Trappe du **Tourteau Departy**, la — land: E of les Monts Heraults, SPB. Tourteau q.v.

Trappe, la — houses, land, fields: les Arguilliers, VdelE; les Hougues, C; les Jaonnets, SSV; les Adams, les Brehauts, le Valangot, le Coudré, SPB; la Croisée, le Colombier, les Rocques, T; les Fontenelles, le Chemin le Roi, Airport, les Bruliaux, F; les Rébouquets, rue des Camps, les Maindonnaux, SM; la Croix au Baillif, SA. Site of megalith, 'chamber tomb'. Déhus, déhuzet, pouquelaie, trépied q.v.

Trappe, rue de la — road: N of les Houmets, VdelE.

Trappe Salmon, la	land: rue des Marais, VdelE. Sallemon q.v.
Traude(s), la(les)	fields: E of les Merriennes, SM. Surname extinct 18th c. **Traude**. Also spelt **Trode**.
Travée, la	land: S of les Laurens, T. 'terrace'. Also spelt **Traouée**, **Traveye**.
Travers, de	fields: les Beaucamps, C; les Villets, F. 'through', i.e. 'crossing'. traversain q.v.
Traversain(s), le(s)	fields: la Couture, St Jacques, les Rohais, SPP; les Vardes, SSP; Bordeaux, CduV; rue des Marais, VdelE; la Hougue Renouf, rue du Presbytère, C; la Claire Mare, rue de l'Arquet, SSV; la Hougue Anthan, les Paysans, les Monts Hérault, SPB; le Portelet, Pleinmont, les Tielles, T; Airport, F; le Vallon, Bon Port, la Quevillette, les Landes, Petit Bot, le Douvre, Jerbourg, le Mont Durand, SM; la Villiaze, SA. Cart track, 'impermanent crossing', without surface or boundaries. Such a track often went across a field in a straight line between entrances, to give access to an inside property (usually a legal servitude). cache, charriere, querriere, q.v.
Trechot, le	see **Truchot**.
Tregale, la	see **Trigale**, la.
Tréhard, le **Camp**	house, land: les Rocquettes, SA. Surname extinct 16th c. **Tréhard**. camp q.v.
Trelade, la	house: la Hougue, SM. Early 20th c. name, now a hotel.
Trembles, la **Vallée** des	land: le Gron, SSV. 'aspen' (poplar).
Trembles, les	land, fields: la Ramée, SPP; les Osmonds, le Courtil de haut, SSP; Le Gron, SSV; le Grais, SPB. Derivation as above.
Tremel, le	field: les Raies, SPB; house: late 19th c. la Grande rue, SM. Origin unknown.
Trencheur, le	field: les Niaux, C. Origin unknown.

Trenquesous, les — field: E of les Prevosts, SSV. Presumably owed 30 sous tournois in rente.

Trenquesous, rue des — road: N of route des Houguets, SSV.

Trépied(s), le(s) — land, fields: St George, les Grantez, la Neuve Maison, C; les Trépieds, le Mont Chinchon, SSV; les Valettes, Fief le Comte, SPB; les Nouettes, F; les Grands Courtils, SA. Site of megalith, 'chamber tomb'. Probably from 'tripod', as used on a hearth. Déhus, déhuzel, pouquelaie, trappe q.v.

Trépied des **Cherfs**, le — land: les Cherfs, C.

Trépied du **Galle**, le — land: S of les Renouards, C.

Trépied des **Nouettes**, le — land: les Nouettes, F.

Trépied, le **Camps** du — land: la Pouquelaye, les Valettes, SPB.

Trésor de St Sauveur le — land: les Hougues, C. Presumably from rente owed to trésor of St Sauveur, i.e. the treasury of St Saviour's church.

Trésor, le — land, fields: les Rohais, rue des Cornets, SPP; les Sauvagées, la Vieille Rue, SSP; les Landelles, C; le Neuf Chemin, SSV; la Pomare, SPB. Presumably owed a rente to the 'treasury' of the parish church.

Trésor St Jean, le — field: le Gain, SPB. Presumably connection with a parish church confraternity of St Jean.

Trésor St Pierre, le — land: les Monts Hérault, SPB. Presumably connection with a parish church confraternity of St Pierre.

Trigale, la — land: le Crolier, T; les Villets, F. Origin unknown

Trinité, la —
1. pre-Reformation dedication of the parish church of Ste Marguerite de la Forêt (Forest parish church).
2. dedication of Trinity Church, la Contrée Mansell, SPP. Clugeonerie q.v.

Triquerie, la — field: N of la Mare de Carteret, C. Origin unknown. First recorded 1734.

Trohardy, de — field: les Ruettes Brayes, SPP. Surname extinct 17th c. **Trohardy**.

Trois Bouvées de l'Eree, Fief des — feudal holding of land. Dependency of Onze Bouvées de Nord-Est du Fief le Comte. SPB.

Trois Cornières, à — see **Cornières**, à **Trois**.

Trois Vatiaux, Fief des — feudal holding of land. Dependency of Fief de Longues, SSV.

Trois Verges, les — see **Verges**, les **Trois**.

Trois Vues, les — see **Vues**, les **Trois**.

Tronches, les — land, fields: Fermains, Havilland vale, SPP; la Hougue Anthan, SPB; les Laurens, T; la Croix Forêt, F; Icart, La Fallaize, SM. 'stumps'.

Tronchières, les — land: les Cherfs, les Queritez, le Marais, C; les Norgiots, SA. Area of 'stumps'.

Tronquesous, le — see **Trenquesous**, les.

Trop Loué, de — fields: Beauvallet, SSV; le Colombier, T. 'too much leased', i.e. too high a rent.

Trop Vendus, les — field: la Grande Rue, SSV. 'too much sold', i.e. too high a sale price.

Troussaillerie, la — land: Bordeaux, CduV. After **Pierre Troussey**, 1654.

Troussey, de — land, fields: St George, C; Petit Bôt, T. Derivation as above. Also feudal holdings of land, Bordage **Troussey**, la Vrangue, SPP; Bordage **Troussey**, S of le Bourg, F.

Truan, au — fields: la Hurette, C. Surname extinct 16th c. **Le Truan**.

Truchon, le — houses: W of les Mares, SPB. Recorded le **Trusson**, 1549- 1762, le **Truchon** from 1792. Origin unknown.

Truchon, le — field: unlocated, SA. Recorded **le Truchon** 1607. Origin unknown.

Truchot(s), le(s) — district: le Truchot, SPP; la Hougue, C; S of les Rohais, SA. Origin unknown.

Truchot, rue du — road: N of le Pollet, SPP.

Trusson, le — see **Truchon**, le.

Turquie, la — houses, land: N of Bordeaux, CduV. Origin unknown. Earliest reference 1755.

Tuzées, les — former house: S of les Eturs, C. see **Déhuzet**, le

U

Usées, **Vusées**, les — land: W of les Plichons, VdelE. Recorded les **Husetz**, 1611. Derived from déhuzet q.v

V

Vache, à la — field: les Monts, SSP; le Bouillon, SA. Possibly after extinct surname, **Le Vacques**.

Vaches, la **rue** des — road: originally between la Grande Rue (High Sreet) and the harbour, SPP, for landing cattle off vessels. Now called Cow Lane.

Vacheul, le — land, fields: les Cocquerels, C; le Courtil Guilbert, SPB; le Vacheul, F; Sausmarez manor, SM. Surname extinct 16th c. **Le Vecheulx**. Vicheris, Vicheux q.v.

Vaindits, les — fields: W of les Rocquettes, SA. Origin unknown. Recorded **Vaindis** 1585.

Val, le — land: les Costils, SPP; la Rivière, C; Saints, SM. 'valley'.

Val, le — houses, land: rue du Val, les Marais, SPB.

	After **Guillaume Du Val**, 1504. Surname extinct 19th c. **Du Val** (extant Jersey). Duval, Duvaux les, q.v.
Val Aubert, **Fontaine**	spring: near les Laurens, T.
Val au **Coulomb**, **Camps** du	land: near le Neuf Chemin, SSV.
Val, de **Rouge**	see **Rougeval**, de.
Val des **Terres**	see **Terres**, les.
Val, le **Mont** de	see **Mont de Val**, le.
Val, le **Mont** le	see **Mont le Val**, le.
Val, le **Rond**	see **Rondeval**, le.

Val, **vau** (pl. **vaux**, **vos**), valley, with proper names attached are listed below. See also individual entries.

Val au Bourg, le	land: Saints, SM	Bourg q.v.
Val au Cocq, le	fields: les Bourgs, SA	Cocq q.v.
Val au Colomb, le	land: E of Reservoir, SSV	Colomb q.v.
Val au Colomb, le	land: le Felcomte, SPB	Colomb q.v.
Val au Febvre, le	fields: les Blicqs, F/SM.	Feyvre q.v.
Val au Machon, le	field: la Neuve Maison, SSV	Machon q.v.
Val au Nô, le	land: S of le Bouet, SPP	Nô q.v.
Val au Vallet, le	field: la Fallaize, SM	Vallet q.v.
Val Aubert, le	land: les Laurens, T	Aubert, q.v.
Val d'Amende, le	field: la Hure Godefrey, SPB	Amende q.v.
Val des Queux, le	land: les Terres, SPP	Queux q.v.
Val des Terres, les	thoroughfare: les Terres, SPP	Terres q.v.
Val Etur, le	fields: la Croisée, T	Etur q.v.
Val Gonde, le	land: Jerbourg, SM	Gonde q.v.
Val Heureux, le	land: les Monts Hérault, SPB	Heureux q.v.
Val Hubert, le	land: les Tielles, T	Hubert q.v.
Val Luce, le	land: le Vallon, SM	Luce q.v.

Val Robert, le	field: la Hure Godfrey, SPB	Robert q.v.
Val Videcocq, le	land: Lefebvre street, SPP	Videcocq q.v.
Valangot, le	fields: les Monts Hérault, SPB	Angot, q.v.
Valmanquin, le	land: E of les Tielles, T	Manquin q.v.
Valniquet, le	fields: Pleinmont, T	Nicet q.v.
Vau Bellée, le	land: les Vauxbelets, SA	Blicqs q.v.
Vau de Dan Philippe, le	douit: les Landelles, C	Dan Philippe q.v.
Vau de Lif, le	fields: la Prévôté, SPB	Lif q.v.
Vau de Monnel, le	field: Pleinmont, T	Monnel q.v.
Vau de Véline, le	land: Pleinmont, T	Véline q.v.
Vau des Epines, le	land: Pleinmont, T	Epines q.v.
Vaupaint, le	field: rue à Gots, C	Paint q.v.
Vaupanier, le	land: E of Queen's road, SPP	Pannier q.v.
Vaupotel, le	fields: la Fosse, SPB	Potel q.v.
Vauquiédor, le	fields: W of rue à l'Or, SPP	Or q.v.
Vaux Bailleul, les	field: les Bémonts, SA	Bailleul q.v.
Vaux Bêtes, les	gulley: Jerbourg, SM	Bêtes q.v.
Vaux de la Croix le	field: Fermains, SPP	Croix q.v.
Vaux Hubert, les	fields: la Fallaize, SM	Hubert q.v.
Vaux Malade, le	field: les Villets, F	Malade q.v.
Vaux Perrons, les	field: les Etibots, SPP	Perron q.v.
Vaux Rougier, les	fields: les Bordages, SSV	Rougier q.v.
Vauxbelets, les	district: les Vauxbelets, SA	Blicqs q.v.
Vauxbourdeaux le	field: le Grand Belle, SA	Bourdeaux q.v.
Vauxlaurens, les	land: St Julian's avenue, SPP	Laurens, q.v.
Voshues, les	fields: les Vauxbelets, SA	Hue q.v.

Six Val(**s**), **vau**(**x**) listed below lack attribution;

Valengeur, le	field: les Ruettes, SA. Recorded **Vallinguellet** 1610, **Valingelet** 1701, **Valinguer** 1793. Possibly from extinct surname **Valinguet**, q.v.
Valinguet, le	house, fields: E of Candie, C; SA. Recorded **Vallinguier** 1585, 1700, **Valinguer** 1746. Possibly from an extinct surname. **Valengeur** q.v.
Valmette, la	field: les Laurens, T. Recorded **Vallemette** 1576. 'valley'. Origin unknown.
Valomos, les	field: les Eperons, SA. Recorded **Valomos** 1723. 'valley'. Origin unknown.
Vauxhanniet, le	land: les Camprots, F. 'sedge valley'. han, hannière q.v.
Vauxpereux, les	field: les Fauconnaires, SA. Recorded **Vospereux** 1610, **Vaupiereux** 1684. 'valley'. Origin unknown.
Vallan, de	land: les Houmets, C. Origin unknown.
Valle, la **Hougue** du	land: Vale Avenue, CduV. The 'Vale hillock', levelled late 19th, early 20th c.
Vallée(**s**), la(les)	houses, land, fields, all parishes. 'valley slope'. Often arranged in terraces.

Vallée au Brullin la	field: les Vallées, C	Brulin q.v.
Vallée au Ferron la	land: les Hougues, C	Ferron q.v.
Vallée au Sermonier la	land: l'Erée, SPB	Sermonier q.v.
Vallée Cateline, la	land: l'Erée, SPB	Cateline q.v.
Vallée de Corbel la	land: les Niaux, C	Corbel q.v.
Vallée de **haut**, la	fields: le Torval, les Hougues, C.	'top slope'.

Vallée de Misère la	land: E of Queen's road, SPP	Misère q.v.
Vallée de Nordest la	land: les Vesquesses, C	Nordest q.v.
Vallée de Rougeval, la	land: Rougeval, T	Rougeval q.v.
Vallée de Vauvert, la	land: Vauvert, SPP	Vauvert q.v.
Vallée des Choffins, la	land: les Choffins, SSV	Choffins q.v.
Vallée des Houlles, la	land: les Effards, C	Houlets, q.v.
Vallée des Nouis la	land: Beauvallet, SSV	Nouis q.v.
Vallée des Trembles, la	land: le Gron, SSV	Trembles q.v.
Vallée du Becquet, la	land: la Fontenelle, SPB	Becquet q.v.
Vallée du Mont Durand, la	land: le Mont Durand, SPP	Durand q.v.
Vallée du Tablet la	land: le Tablet, SPB	Tablet q.v.
Vallée Enoc Le Patourel, la	land: le Camp Tréhard, SA	Patourel q.v.
Vallée Genâts, la	land: les Niaux, SA	Genâts q.v.
Vallée Guillemotte, la	land: l'Echèlle, SA	Guillemotte q.v.
Vallée Jenas, la	land: le Monnaie, SA	Genâts q.v.
Vallée Jénémie, la	land: la Pomare, SPB	Jénémie, q.v.
Vallée Renault, la	fleld: les Cambrées, SPB	Renault q.v.
Vallée Robert Daniel, la	land: le Pré, SPB	Daniel q.v.
Vallée Simon, la	fields: les Clercs, SPB	Simon q.v.

Vallée Thomellin la	land: les Hèches, SPB	Thomellin q.v.
Vallée, rue de	road: unlocated, SM	
Vallée, **Clos** de la	land: le Douit, C	
Vallée, la **Verte**	land: S of les Laurens, T	'green slope'.
Vallées de Quanteraine, les	land: les Vallettes, SPB	Quanteraine q.v.
Vallées Pipet, les	fields: Candie, C	Pipet q.v.
Vallées, les **Basses**	land: le Gouffre, F	'bottom slopes'.
Vallées, les **Hautes**	fields: E of le Groignet, C	'top slopes'.
Vallées, les **Petites**	land: E of le Groignet, C	'small slopes, terraces'.
Vallées, **ruette** des	lane: S of le Préel, C.	

Vallen, de — see **Vallan**.

Vallen, les **Camps** — land: l'Ardaine, SPB. Origin unknown.

Vallet, le — houses, fields, all parishes. Diminutive of val, valley. 'small' valley. Also spelt **Valet**.

Vallet, **ruette** du — land: W of rue du Vallet, SPB.

Vallet, le **Val** au — field: la Fallaize, SM.

Vallette, la — houses, land: Havelet, SPP; les Hures, CduV; les Vallettes, SSV; les Sages, la Hougue Anthan, St Briocq, SPB; les Tielles, T. Diminutive of **vallée**, 'small' area of slope. Also spelt **valette**.

Vallettes Ingot, les — land: le Gain SPB. Extinct surname **Ingot**.

Vallon, le — houses, land: Rouge Rue, SPP; les Genats, Candie, les Fontaines, C; le Bigard, F. a 'hollow'.

Vallon, le — house, land: le Hurel, SM. Name given in 19th c. to rebuilt gentleman's residence (General de Vic Carey).

Valmette, la **Hure** — land: near les Laurens, T.

Valnord, de — houses: Collings road, le Mont Durant, SPP. Name of two 19th c. villas in above areas.

Valpied, de — fields: les Prevosts, Beauvalet, SSV. Extant surname **Valpied**. Also spelt **Vallepy**, **Vallepied**.

Vannier, la **Ville** — see **Ville Vannier**, le.

Vara, de — see **Varivaray**.

Varclin, de — houses, land: E of les Maindonnaux, SM. Recorded **Varclin** 1511. Origin unknown.

Varclin, **Fontaine** — spring: le Varclin, SM.

Varde Guille, la — field: les Vardes, SPP. Extant surname **Guille**.

Varde(**s**), la(les) — houses, land: les Vardes, SPP; l'Ancresse, CduV; l'Erée, SPB; Pleinmont, T. 'steep sided height'.

Varde, la **Platte** — field: les Vardes, SSP. 'flat' (on top of the height).

Vardes, la **Moulin à vent** des — windmill: les Vardes, SPP. 18th c. construction. moulin q.v.

Varende(**s**), la(les) — see **Garenne**.

Varendes, route des — road: E of l'Aumone, C.

Varinvarays, rue des — road: unlocated, SM.

Variouf, **Boissellée de Henry de**, Fief de la — feudal holding of land. Dependency of Fief Bequepee, SPB.

Variouf, le — houses, land: le Variouf, F; les Ruettes, SA. Surname, **De Vaurouf** extinct 15th c. **Du Vauriouf**, extinct 17th c. **Jean De Vaurouf**, 1511, heirs **Pierre Vauriouf**, 1654. Varouf, Vaurioufs q.v.

Variouvez, les — land: la Planque, les Villets, F. Recorded les **Varyouvées**, 1562. Plural of extinct surname **De Variouf** q.v.

Varivarays, les — fields: S of les Ruettes Braye, SM. Contracted to **Vara**, 20th c. Probably from an extinct surname.

Varouf, au	land: la Bouvée, SM. Surname **De Vaurouf**, extinct 15th c. **Du Vauriouf**, extinct 17th c. Jean **De Vaurouf**, 1511, heirs **Pierre Vauriouf**, 1654. Varouf, Vaurioufs q.v.
Varouf, le **Camp**	land: l'Erée, SPB. Derivation as above.
Varouf, le **Mont**	land: les Padins, SSV. Derivation as above.
Varvatière, la	house: la Houguette, SPB. 19th c. nickname, smelly dung heap.
Vassal, la **Chasse**	road: Cimetière des Frères (q.v.) to marker stone for les Barrières de la Ville (q.v.) at rue des Forges (Smith Street, q.v.). i.e. for most part; Ann's Place, SPP. Surname extinct 15th c. **Vassal**. Recorded as **la Cache Vassal**, 15th & 16th centuries. fontaine q.v.
Vassalerie, la	house, land: Ste Hélène, SA. Derivation as above.
Vasse, ruette	see **Thomas Henry**, la **ruette**. SPB.
Vatines, les	fields: la Vieille Rue, SSV. Recorded les **Vatines**, 1695. Origin unknown.
Vatre, de	field: les Etibots, VdelE. Recorded les **Vattres**, 1749. Origin unknown.
Vau(x), le(s)	see **Val**.
Vauclins, les	fields: N of le Creux Mahié, T. Surname extinct 16th c. **Vauquelin**. Spelt **Vauxclins** 19th & 20th centuries.
Vaucourt, de	fields: Candie, C; la Fallaise, Calais, Saints, SM. Surname extinct 19th c. **Vaucourt**.
Vaudin, de **Toussaint**	fields: les Longs Camps, SSP; les Landes du Marché, VdelE. After **Toussaint Vaudin**, 16th c.
Vaudin, la **Hougue**	unlocated: SSP. Derivation as above.
Vaudin, le **Clos**	field: N of rue a l'Or, SPP. Extant surname **Vaudin**.

Vaudinerie, la — houses, land: Victoria road (la Couperderie), SPP. Derivation as above, from surname **Vaudin**.

Vaudines, les — field: l'Echelle, SA. Feminised form of extant surname **Vaudin**, i.e. field had been in a woman's name for a long time.

Vaugrat, le — district: Vingtaine de Nermont, SSP. Contracted form of surname extinct 14th c. **De Vaugerard**. Grat, le port q.v. Also feudal holding of land, Fief de **Henri de Vaugrat**. **Henry de Vaugerard**, 1309.

Vaugrat, route de — road: S of route de Port Grat, SSP.

Vauquelins, les — see **Vauxclins**, les.

Vauquiedor, le — land: to N of present Princess Elizabeth Hospital, SPP/SA. 'valley which is of gold', i.e. colour of the valley gravel. Or, le Vauquiéd' q.v.

Vaurioufs, les — house, land: les Loriers, CduV; les Vaurioufs, la Bouvee, la Fallaize, SM. Surname, **De Vaurouf** extinct 15th c, **Vauriouf**, extinct 17th c. **Jean De Vaurouf**, 1511, heirs **Pierre Vauriouf**, 1654. Also feudal holding of land, Fief de la **Boisselée de Henry du Vauriouf**, SSV. (12th c.). Varioufs, Varouf, q.v.

Vautier, de — field: N of Fermains, SPP. Surname extinct 15th c. **Vautier** (extant Jersey).

Vautiers, le **Clos** — field: E of le Gelé (S of la Grande Mare), C. Derivation as above. Also feudal holding of land, Bouvée **Vautier**, C.

Vauvert, de — district: Vauvert, SPP. Surname extinct 17th c. **Vauvert**. Heirs **James Vauvert**, SPP, 1574.

Vauvert, la **Hougue** — land: near le Platon, SPP.

Vauxlorens, **ruette** des — land: S of les Cotils, SPP.

Vazon, de — 'large area' of beach.

Veaux, les — fields: le Marais, la Ronde Cheminée, SSP. From **Vaux**, plural form of surname **Du Val**.

Veaux, le **Clos** au — field: le Hèchet, C. Derivation as above. clos q.v.

Vèche, à — fields: la Lande, CduV; les Messuriers, SPB. 'vetch'. Recorded **à Vesse**, 1595.

Vecheulx, le — see **Vacheul**, **Vicheris**, **Vicheux**.

Vecque, le — fields: le Clos Vivier, SSV; N of les Nouettes, F. Alternative spelling of **Vèche**. Spelt **Veque** since 18th c. Veque, Vesquesses q.v.

Vée, le **Camp** au — field: les Huriaux, SSV. Extinct surname, **Le Vée** (extant Jersey). camp q.v.

Végétaux, les — field: les Houards, F. 'vegetables'. 18th c. name.

Vehue, la — fields: W of les Villets, F. 'Hue's way', referring to the road alongside. Recorded **la Voye hue**, 1671.

Veleresse du Côté de la Fallaize, Fief de la — feudal holding of land. SM

Veleresse du Côté de Fermains, Fief de la — feudal holding of land. SM.

Veline, le **Vau** de — land: le Portelet, T. 'wild mustard'. val q.v.

Venelles, la **ruette** des — land: la Couture, SPP.

Venelles, les — field: la Couture, SPP. At 'the alleys'.

Ventelles, de — field: le Hurel, CduV. 'made known', i.e. for lease or sale. Recorded de **ventelaie**, 1591.

Vents, les **Quatre** — house, land: les Huriaux, SM. 19th c. house name, earliest reference 1861.

Veque, le — see **Vecque**.

Vequesse(s), la(les) — see **Vesquesses**.

Verbeuse, la — land: les Camps. SM. 'grass verge', from herbeuse. Recorded **verbeuse**, 1511.

Verbeuse Dom Hue	land: le Courtillet, SM. Dan Hue q.v.
Verbeuse Ferron	land: le Coin Fallaize, SM. Ferron q.v.
Verbeuse Robinet	land: les Camps, SM. Robinet q.v.
Vere, la **Rocque au**	land: Bon Port, SM. Derivation as above. rocque q.v.
Vere, le **Camp** au	land: Bon Port, SM. From an extinct surname. Recorded **Campauvere** since 18th c. camp q.v.
Verges, les **Cinq**	fields: W of la Vieille rue, SSV; Pleinmont, T. Literally strips of land 5 yards in width.
Verges, les **Trois**	fields: le Coudre, SPB; le Courtillet, SM. Literally strips of land 3 yards in width.
Vergée(s), la(les)	land, fields: la Grosse Hougue, SSP; la Rochelle, les Coutures, CduV; la Pouquelah, C; l'Erée, SPB. Former open areas in camps de terre, originally split into strips of a 'vergée' of land (square measure). Also spelt **vergie**.
Vergies, rue des	road: unlocated, C.
Vermeuleurs, les	fields: E of les Huriaux, SA; SM. Origin unknown. Recorded **Vermelus**, 1685.
Verron, le **Clos**	fields: l'Aumone, C. From surname Verron. clos q.v.
Verte Rue, la	see Rue, la **Verte**.
Vésiez, au	field: Woodlands, C. Surname extinct 15th c. **Le Vesiez**.
Vésiez, le **Camp** au	land: les Queritez, C. Also spelt **Vesies**, **Vezies**, **Vessies**.
Vesquesses(s), la(les)	house, land: le Dos d'Ane, C; les Blanches Pierres, SM. 'vetch patch'. Recorded **Vequeces**, 1536.
Vibert, de	field: St Clair, SSP. Surname extinct 16th c. **Vibert**. (extant Jersey).
Vic(q), **De**	fields: le Marais, SSP; les Blicqs, SA.

	Surname extinct 17th c. **De Vic**, **Devicq**.
Vic, les **Messières Thomas De**	land: le Mont Crevelt, SSP. Derivation as above. Also feudal holding of land, Fief de **Robert De Vicq**, SPB/T.
Vicheris, les	fields: la Pomare, SPB. Surname extinct 16th c. **Le Vecheulx**.
Vichet, le **Clos**	field: S of la Grande Mare, C. Surname extinct 15th c. **Vichet**.
Vicheux, la **Mare** au	land: S of les Sablons, SPB. Originally la **Mare** au **Vecheulx**. Mare, Vacheul q.v.
Victoria Tower	see **Hyvreuse**, l'.
Vidamours, de	field: la Belle, SPB; le Bout, T. Extant surname **Vidamour**.
Vidamours, de **Collas**	field: l'Alleur, T. Recorded 1595. Derivation as above.
Vidamours, de **Helier**	field: le Marais, C. Derivation as above.
Vidamours, de **Jean**	field: le Courtil Guilbert, SPB. Derivation as above.
Vidamours, de **Moise**	field: les Corbinez, SPB. Derivation as above.
Videclin(s), les	house, fields: les Videclins, C; les Huriaux, SA. Surname extinct 15th c. **Viteclin**.
Videclins, le **Camp** des	land: les Pelleys, C. Derivation as above. Also feudal holding of land, Fief des **Videclins**, C.
Videcocq, le	house, land: N of route du Braye, CduV; S of les Falles, SPB. Surname extinct 15th c. **Videcocq**.
Videcocq, le **Val**	land: Lefebvre street, SPP. Derivation as above.
Videcocq, rue du	road: E of les Mares, SPB. Error in location made by O.S. surveyors, 1898. It is historically at les Falles, SPB, a mile away.
Vieillard, le	fields: N of les Grands Moulins, C. Surname

	extinct 15th c. **Le Vieillart**. Also feudal holding of land, Bouvée au **Vieillard**, T.
Vieille Maison, la	see **Maison**, la **Vieille**
Vieille Rue, la	see **Rue**, la **Vieille**.
Vieilles, la, les,	1. field: la Lande, CduV. Recorded les **vieilles rues**, 1654. 'old roads', i.e. established ways. 2. fields: la Hougue Fouque, SSV; les Monts Héraults, SPB; les Blanches Pierres, SM. 'of the old (woman)'.
Viel Mont, le	see **Mont**, le **Viel**.
Vieux Courtil le	**see Courtil, le Vieux**
Vieux Hautgard, le	see **Hautgard**, le **Vieux**
Vieux Ménages, les	see **Ménages**, les **Vieux**.
Viétie, le **Clos** au	field: les Pins, C. Extinct surname **Le Viettu**.
Villain, de **Roger Le**	land: la Fallaize, SM.
Villains, les	field: le Crolier, T. Surname extinct 15th c. **Le Villain**.
Ville, à **Haute**	see **Hauteville**.
Ville ès **Duriaux**, **Fontaine**	spring: unlocated, SPP.
Ville, la	1. town: i.e. la **Ville** de **St Pierre Port** 2. an 'isolated holding or settlement' dating from the medieval period.
Ville, la **Neuve**	1. 'New Town': district N of le Truchot, Chapelle St Julien, SPP. Recorded **Neuve Ville** 1574. 2. 'New Town': district, Sausmarez street, Havilland street, George street and Union street, developed 1800–1840. Recorded **Neuve Ville** 1810.
Villette, la	houses, land: la Villette, SM. 'small' settlement. Possibly a 'villata', hamlet. Recorded 13th c. 'franchises', Extente 1331.

Villiaze, la	houses, land: la Villiaze S of Airport, F. The ending **Ase**, **Aze**, would appear to predate the regular hereditary surnames of the later 12th c. Recorded 1671.
Villiaze, la	houses, land: la Villiaze N of Airport, SA. Derivation as above. Recorded 1592.
Villocq, le	houses, land, fields: le Villocq, C. **Oc** attached would appear as above to predate the regular hereditary surnames of the later 12th c. Recorded 1508. Also spelt **Villiocq**. croix q.v.
Villocq, le	field: with modern house, les Adams, SPB. Derivation as above. Recorded 1671.
Villocq, douit de la	field: le Villocq, C.
Villocq, la rue(tte) du	land: W of le Villocq, C.
Villory, la	field: les Arguilliers, VdelE. **Ori**, **Ory** attached would appear as above to predate the regular hereditary surnames of the later 12th c. Recorded 1583.
Ville Amphrey, la	houses, land, road: S of les Camps. SM. Recorded 1624, viz. **Hamfroy**, 1624, **Anfray** 1661, **Anfre** 1676, **Enfre** 1711, **Amphre** 1725. Surname extinct 16th c. **Enfraye**.
Ville Baudu, la	houses, land: E of Vale church, CduV. Extinct surname **Baudu**, **Bodu**. Recorded 1309.
Ville Bichard, la	see **Ville Pichart**.
Ville à Beuf, la	fields: rue des Bergers, C. Surname extinct 16th c. **Le Boeuf**. Recorded 1624. Beuf, le q.v.
Ville Dan Hue, la	field: W of le Clos Vivier, SSV. Recorded 1555. **'Dominus Hue'**, **Dan Hue**. domaine, fief, moulin à vent, verbeuse, q.v.
Ville aux Duraux, la	site: S of les Damouettes, SPP. Surname extinct 15th c. **Durel**. Recorded 1571. Durel q.v.

Ville Hérode, la	fields: route de Plaisance, SPB. Extinct surname **Hérodes**. Recorded 1504.
Ville ès Patris, la	site: unlocated, rue des Landes, SM. **Pierre Patris** owned la **Ville ez Patris**, SM, 1517. Surname extinct 16th c. **Patris**. Patris q.v.
Ville au Petit, la	land: S of le Vauquiédor & E of the PEH, SPP. Recorded 1574. Surname extinct 17th c. **Le Petit**. The form **Petit** remains extant.
Ville Pichart, la	1. land: E of rue d'Enfer, S of la Grande Mare, C. Recorded 1507. Surname extinct 15th c. **Pichart**. 2. former field: now part of le Clos Théhy q.v., opposite Seaview farm, C.
Ville ès Pies, la	land: N of Belval, CduV. Surname extinct 17th c. **La Pie**. Recorded 1632. Pies q.v.
Ville ès Riches, la	field: unlocated, CduV. Surname extinct 15th c. **Le Riche** (extant Jersey) Reference 1457.
Ville au Roi, la	house (now demolished), land: la Ville au Roi, SPP. Recorded 1574. Presumably from extant surname **Le Roi**, **Le Roy**. (No evidence for legend re Gaultier de la Salle, 1320.)
Ville Salmon, la	field: les Taudevins, SPB. Recorded **Sallemon**, 1671. Extant surname **Salmon**.
Ville Sauvey, la	field: N of Saumarez Park, C. Located in field behind dwellings NNE of Saumarez park. Recorded 1551. Surname extinct 15th c. **Sauvey**.
Ville Sayve, la	site: E of les Douvres, SM. 'Ville de Seve' from forfaiture of **Richard Sewe**, 1274. Recorded 'at end of rue du Fief au Bret' (rue des Aubrets), 1503, 1579. Spelt **Seyve**, **Sayne**, **Sceues**, **Seyvesez**, **Sevez**. Sevée q.v.
Villetole, la	houses, land: la Lague, T. Earliest reference 1517. Surname extinct 14th c. **Tolle**, **Toll**.

Ville Vannier, la — land: le Hamel, VdelE. Presumably from surname extinct 16th c. **Vannier**. Earliest documentary reference 1885.

Villets, les — houses, land: les Villets, F. Recorded 1562. Surname extinct 14th c. **De Ville, De Villez**.

Vimiéra — property: les Rohais, SPP. Former French monastic house, l'Institut des Freres des Ecoles Chrétiens, built at end of 19th c. and demolished 1980s. Replaced with St Pierre Park hotel.

Vinaires, les — house, land, fields: les Rohais, le Pont Renier, SPP; les Vinaires, SPB. Surname extinct 17th c. **De Vinaire, Vinaire**.

Vingt et Unième Boisselée, Fief de la — feudal holding of land. Dependency of Fief de Rozel. VdelE.

Vingtaine, la —

1. detached division of a parish –
   Vingtaine de Noirmont, St Sampson
   Vingtaine de l'Epine, Vale
   Pleinmont, Torteval
2. minor administrative division within the parish, originally based on the concept of 20 households, more usually called canton during the 20th c. The historic vingtaines (cantons) of SPB are les **Yvelins**, les **Marchez, Rocquaine** and les **Adams**. Several parishes have lost their historic vingtaines and replaced them with cantons. Epine, Noirmont, Pleinmont q.v.

Vingtième, le — fields: la Maison de haut, SSV; la Pomare, SPB; Saints, SM. 'one twentieth', which related to the former additional share inherited by the eldest son in a partage des immeubles (division of real property). Abolished in 1954.

Vingt Maisons, les — see **Maisons**, les **Vingt**.

Vivemer, de — land: les terres, SPP; les Houmets, C. Surname extinct 15th c. **Vivemer**. **Jehannet Vyvemer**, 1409. Also feudal holding of land, Bouvée **Vivemer**, C.

Vivette, la — field: les Hougues, C. 'poor living'.

Vivier, le — man made 'ponds': la Ville au Roi, les Granges de Beauvoir, la Vrangue, SPP; les Annevilles, SSP; le Groignet, C; les Lohiers, SSV; la Mare Hailla, SPB; Sausmarez manor, SM; les Grands Courtils, les Vauxbelets, les Rohais, SA. mares q.v.

Vivier, le **Clos** — land, fields: le Clos Vivier (le Rocher), C; le Clos Vivier (le Gron) SSV. Donation of lands above, detailed, by **Godefri Vivier** to abbey of Mont St Michel, before 1179. (Cartulaire des Iles Normands, p.219)

Voies, les **Trois** — see **Vues**, les **Trois**.

Voie, la **Croix de Belle** — see **Belle Voie**, la **Croix** de.

Voie Cornillot, la — way: unlocated, le Gain, SPB. Surname extinct 15th c. **Cornillot**.

Voie de l'Eglise — see **Vue** de **l'Eglise**.

Vrangue, la — houses, land, fields: la Vrangue, SPP; la Vrangue (la Planque), T; la Vrangue (les Villets), F. Extinct surname of probably 12th c. but record lost.

Vrangue, les **Moulins à eau** de la — upper and lower water mills: la Vrangue, SPP. Water course and some ruins of the lower mill remain. Moulins à eau q.v.

Vrée, la **Croûte** au — house, field: les Clospains, CduV. 'varaic, varech'. seaweed.

Vrée, la rue au — road: les Clospains, CduV. Derivation as above. Earliest reference **Vrec**, 1808.

Vue de l'**Eglise**, la — fields: W of Forest church, F. 'church view'. Originally **Voie** de l'**Eglise**, 1671, 'church way', and road adjacent known as rue des **Grons** q.v.

Vues, les **Trois** — road junction: les Rohais (le Foulon), SPP. 'three views'. Originally les **Trois Voies** 1574, 'three ways'.

Vusées, les — see **Usées**.

W

Well road — see **Puits**, la **rue** du.

Y

Yvelin, le **Camp** — field: les Issues, SSV. Surname extinct 15th c. **Yvelin**.

Yvelin, la **Croix** — field: E of les Brehauts, SPB. croix q.v.

Yvelin, les **Landes** — fields: les Landes, S of above Croix, SPB.

Ysorey, d' — see **Orais**, les

Ysorey, la rue d' — now Route de Côbo.